十年十划

赵春水 著

江苏凤凰科学技术出版社·南京

图书在版编目（CIP）数据

十年十划 / 赵春水著 . -- 南京：江苏凤凰科学技术出版社 , 2022.10
 ISBN 978-7-5713-2778-1

Ⅰ . ①十… Ⅱ . ①赵… Ⅲ . ①城市规划 – 建筑设计 – 研究 – 天津 Ⅳ . ① TU984.221

中国版本图书馆 CIP 数据核字 (2022) 第 028635 号

十年十划

著　　　者	赵春水
译　　　者	刘晓烨　陈　旭　张　萌
项 目 策 划	凤凰空间 / 陈　景
责 任 编 辑	赵　研　刘屹立
特 约 编 辑	狄　阙　田　园
出 版 发 行	江苏凤凰科学技术出版社
出版社地址	南京市湖南路 1 号 A 楼，邮编：210009
出版社网址	http://www.pspress.cn
总 　经　 销	天津凤凰空间文化传媒有限公司
总经销网址	http://www.ifengspace.cn
印　　　刷	天津图文方嘉印刷有限公司
开　　　本	710 mm×1000 mm　1 / 16
印　　　张	13
字　　　数	208 000
版　　　次	2022 年 10 月第 1 版
印　　　次	2022 年 10 月第 1 次印刷
标 准 书 号	ISBN 978-7-5713-2778-1
定　　　价	98.00 元

图书如有印装质量问题，可随时向销售部调换（电话：022-87893668）。

前言

2020年初，突如其来的"新冠肺炎疫情"打乱了我的一些计划，《十年十划》的出版就是被中断的工作之一。原计划对2009—2018年完成并实施的城市规划设计项目做一个阶段性总结，该工作于2018年启动，预计2020年底或转年初完成，但不得不推迟至今日。

本书所载项目是从完成的百余个规划成果中精选的10个代表作品，为方便读者阅读，将项目分为三类：公共建筑集群类、住宅社区类和城市更新类。通过同类项目之间的相互比较，揭示该类项目在规划理论、技术方法上的异同，剖析项目背后顺应时代发展的演化历程，进而为应对不断变化的需求，归纳出合理适宜的技术路径和规划策略。

一是公共建筑集群类，主要就近年参与或主持的三个文化中心项目进行论述，聚焦文化中心类项目的迭代和演化，由注重礼仪性、地域性和秩序性的天津文化中心（1.0），到重视集约、活力和复合性的天津滨海文化中心（2.0），再到以生态环境优先，集文、商、旅为一体的济宁文化中心（3.0），主要就文化中心项目的递进演化历程、创新动力及设计对策进行对照分析和总结归纳。

天津文化中心（2009—2012年）依天津市区总体开放空间规划进行选址，形成从水西公园、水上公园、天塔地区、银河广场至海河的东西向城市开放空间。文化中心采取分散布局方式，文化建筑集群融入公园环境，延续天津因水而兴、逐水而居的发展轨迹，围绕中心湖组织"海棠轴线"与行政中心呼应。同时，天津文化中心注重城市礼仪性和庄重感的营造，将山、水、塔等传统园林格局活用，构建具有城市象征意义的文化家园、城市客厅，成为反映时代特征的公共空间典范。

同天津文化中心强调礼仪性、地域性和秩序性不同，天津滨海文化中心（2013—2018年）的规划务实而不平庸，规避了大型城市广场带来的空泛与贫瘠，采取收缩集中布局增加商业份额的策略，探索文商融合的建设模式。以文化长廊集聚各个文化建筑，形成文化街道的空间范式，文化搭台、商业唱戏，吸收天津文化中心设置能源中心的经验，统筹能源、交通打造绿色发展模式。文化长廊将传统意义的文化中心转化为以长廊空间为核心的文化建筑综合体，同时长廊承载了文化建筑外溢的商业需求，它的存在标志着新文化建筑类型的诞生。

继天津文化中心、天津滨海文化中心之后，济宁文化中心（2014—2020年）

是我们的团队独立完成的城市文化中心规划设计项目。济宁有"孔孟之乡""运河之都"的美誉，项目选址位于济宁市西南太白湖东岸，自然资源丰厚，传统文化礼序影响深远。济宁文化中心规划以传承"孝天敬人"的古训为主旨，将尊重传统人文礼序转化为遵守自然生态环境秩序，以太白湖自然要素融入文化中心为优先考虑因素，统筹文化平台、单体建筑、外部景观、内部空间、交通市政、能源中心，践行绿色发展理念，将文化、商业、旅游等要素融合，创造展现济宁悠久文化，面向未来的可持续发展新形象。

二是住宅社区类，包括践行"新规划标准"（2018）先行先试的天津实践"河西八大里"项目，生态环境驱动的北辰双青新家园保障房项目，保护与开发协调的"海河院子"天津市第一热电厂项目，以及历时十余年反复优化与时俱进的西站西于庄城市设计项目。

河西八大里（2013—2018年）是天津市政府主导的区域开发项目，又被称为新八大里。2018年实施的《城市居住区规划设计标准》倡导以人为本的原则，推行以使用者行为为标准的规划理念，从制度上化解现行规范与城市发展的矛盾，为科学规划提供了基本制度保障。新八大里规划用地约200 hm^2，借鉴先进规划理念，落地实施了"窄路密网""开放街区""混合住区"，统筹地上、地下资源，在规划指标综合平衡方面创新性地解决了新规划要求与当时规范的冲突，保障了规划实施的完整性和系统性。难能可贵的是，新八大里项目的规划及实施是在2013年前后，那时新标准尚未发布，它有幸成为新标准出台之前城市规划创新的天津实践。

双青新家园（2012—2018年）是天津开展大规模保障性住房规划建设的项目，选址在北辰西部双青地区。场地南北方向有两条次干道连接周边，东西方向被铁路及河道限制，边界有三条二级河道连通，内部水系丰富，植物覆盖率高。规划体现生态优先、以人为本的理念，提出"两河一谷、生态人居"的规划愿景，利用内部地形及高差构筑连通东西河流的"生态绿谷"，串联各种功能形成复合型城市开放空间，使双青新家园成为空间环境品质优良的保障房项目。

天津市第一热电厂项目（2014—2023年）始于海河开发，自2004年海河改造启动了天津新一轮发展的引擎，围绕海河发生了巨大及持续的变化，其中沿岸原租界因其特殊价值得到了很好的保护，但老厂房却没有那么幸运。我们接手该项目的时候，第一热电厂标志性的大烟囱已经被爆破拆除，只剩下主厂房。我们介入的第一件事就是从海河开发整体思考沿岸开放空间的连续性并关注稀缺性，规划通过保护老厂房及其沿河开放空间，围绕老厂房开展构想，提出"海河院子"的规划愿景，沿海河打造体现城市发展轨迹、面向未来的优质河岸生活空间。

西站西于庄（2010—2019年）位于南运河北侧与海河交汇处，南运河北岸是天津市内高铁枢纽西站。西于庄是当时天津最大的棚户区之一，2013年启动拆迁工作的同时开展新一轮规划设计。当时《天津市城市总体规划》正值修编阶段，经论证将该地区发展定位为天津城市副中心，初步形成以小白楼地区为主中心、西于庄地区和柳林地区为副中心的天津未来城市发展格局。于是，在西于庄开展了以中央商务区（CBD）为导向的城市开发规划工作，并形成最终方案，但由于拆迁的迟滞未实施。直到2019年拆迁工作终于完成，但时过境迁，加之京津冀的协同发展上升为国家战略，生态优先、可持续绿色发展理念对原规划提出新的要求，设计团队结合实际与时俱进，提出用中央活动区（CAZ）代替CBD的定位，以务实、可行的方案为西于庄地区发展重新量身定制规划愿景。

三是城市更新类，有历时十几年持续深入优化，不断提升完善的西开教堂周边改造项目；有基于保护历史街区风貌，恢复区域活力的泰安道五大院项目；还有坚持多方位、低冲击、微改造、渐进式有机更新的义乌异国情街的改造项目。

西开教堂周边的规划（2008年至今），该教堂位于和平区滨江道独山路原墙子河外老西开一带，今营口道与西宁道交口。1916年由法国传教士修建。教堂周边聚集西开小学、约瑟小学、约瑟会修女院法汉学校等教育建筑，也是天津重点保护历史风貌建筑。由于历史原因，周边环境恶化，西开教堂在空间上陷入拥挤杂乱的夹缝之中，其艺术价值及历史地位逐渐丧失。书中记录了教堂周边逐步演进的环境提升方案，也反映了各个时期人们对历史建筑保护工作的认识在不断进步。虽然方案没有全部实施，但对周边环境的保护及营造保护历史街区的社会氛围起到重要作用，更为今后历史建筑的保护发展争取了时间，保留了空间。

泰安道五大院项目（2008—2015年），位于海河沿线，北接津湾广场，南接小白楼商圈直至五大道历史街区。2009年天津市政府为恢复历史街区、保护文化遗产，启动以泰安道维多利亚花园（今解放北园）为核心的城市更新建设。首先从区位出发，对津湾广场、泰安道地区、小白楼地区、五大道地区进行统筹规划，谋划沿海河至五大道的发展带，为历史街区复兴带来巨大发展机会。区域内有 16 处历史文物保护建筑，但其分散的布局没有形成整体优势，规划采取重塑街区、链接空间、强化步行等方式重新组织街巷空间，通过院落设置使新老建筑成为有机整体，同时引入新业态丰富商业类型，形成"以保护历史街区为导向，以重塑街区尺度为方式，以恢复步行街巷生活为目标"的天津历史街区城市更新范式。

义乌异国风情街项目（2014—2019 年）源自泰安道五大院的成功，甲方慕名而来，指名邀请参加过五大院项目的团队参与规划设计。面对新的挑战，我们的团队没有先入为主地植入概念，而是以调研为基础逐步开展工作。经过征求群众意见、听取当地政府要求，确定了更新原则，即"尊重居民意愿、尊重市场规律、尊重城市历史"，结合社会管理、经济结构、街区环境，提出"三变三不变"的指导方针，将更新内涵与保持稳定，同时注入活力的核心，通过交流与居民形成共识，形成多方位、有创意、低冲击、微改造、渐进有机更新的义乌模式。

本书的目标不是通过介绍建成效果来展示项目，而是通过重新思考每一个项目，来梳理连续性工作背后的深层逻辑，以期让城市规划远离偏见回归其本质。另外，本书不是一本城市规划理论书籍或通过实际案例来进行理论总结，而是更注重厘清每个项目的来龙去脉，还原真实的情境，记录在当时历史时期的客观条件下，对现实问题的主动回应及深入思考后的解决方法。

一个时期有一个时期的需求，《十年十划》反映当时的需求，记录着时代大潮之下城市发展的轨迹。十年在城市历史进程中虽然不长，但这段时期天津的城市发展取得了很大成就，也留下了不少遗憾，需要我们冷静地反思和真诚地面对，城市就像生命体一样需要生长，现阶段任何褒奖与贬斥都为时尚早。希望本书记录的规划设计历程与思考，能为应对未来更多挑战提供参考与借鉴。

赵春水
2022 年 6 月于兰坪路

目录

天津文化中心　　　　　　　　　　　　　　　　8
Tianjin Cultural Center

天津滨海文化中心　　　　　　　　　　　　　28
Tianjin Binhai Cultural Center

济宁文化中心　　　　　　　　　　　　　　　48
Jining Cultural Center

河西八大里　　　　　　　　　　　　　　　　74
Badali Community, Hexi District

双青新家园　　　　　　　　　　　　　　　　94
Shuangqing New District

义乌异国风情街　　　　　　　　　　　　　　112
Yiwu Exotic Street

天津市第一热电厂　　　　　　　　　　　　　132
Tianjin No.1 Thermal Power Plant

西站西于庄　　　　　　　　　　　　　　　　150
Xiyuzhuang Area of Tianjin West Railway Station Business District

西开教堂　　　　　　　　　　　　　　　　　170
Xikai Church

泰安道五大院　　　　　　　　　　　　　　　186
Five Courtyards in Tai'an Road

设计团队名单　　　　　　　　　　　　　　　206
Team Members

位于渤海之滨、九河下梢的天津卫，水量丰沛，沼泽密布。从天子渡口到建卫至今，600多年间人们都将其视为承载梦想的乐园。民国时期人们从世界各地集聚津沽，在此营造理想的生活环境，留下了现称为五大道、意式风情区、老城厢等汇聚众人智慧的城市人居场所。因河流密布，城市逐水而建，因水而兴。进入21世纪初，海河成为天津城市发展的引擎，海河两岸改造拉开城市发展序幕。随着环渤海经济圈的确立，天津市政府决定建设天津文化中心展现改革开放的发展成果和"文化强市"的雄心壮志。

天津文化是典型的"华洋杂糅"北方文化，回顾天津近代百年的历史可以全面真实地体会到天津曾经的沧桑与辉煌。海纳百川的地域文化让天津在20世纪就成为开放、包容、时尚的先进之地。秉承天津的地域文化基因，文化中心的布局以开放的轴线将新建区与建成区有机地整合起来，辅以景观湿地，堆山筑塔，使整个场地成为兼具城市礼仪性和本土市民文化亲和力的城市广场。

以场地原有水面为中心，文化建筑分别临水而筑，南侧硬朗笔直的岸线结合雕塑布置博物馆、美术馆、图书馆；

天津文化中心
Tianjin Cultural Center

用地面积：90 hm²
建筑规模：1 000 000 m²

城市设计团队：
天津市城市规划设计研究总院

修建性详细规划团队：
天津市城市规划设计研究总院

获奖情况：
2019年度全国优秀勘察设计项目
2014年度第十二届中国土木工程詹天佑奖
2014年度天津市科学技术进步奖一等奖
2013年度全国优秀城乡规划设计奖一等奖
2013年度中国文化建筑优秀工程
2013年度天津市优秀城乡规划设计奖一等奖

鸟瞰图 | aerial view

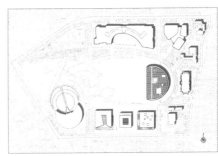

区位图 | location

北侧舒展微曲的岸线结合场地布置商业中心；西侧人工湿地草坡入水，堆山理水布置绿色屏障；东侧以大剧院为底景，营造大型人工亲水平台承载市民亲水而居的梦想，亦成为面向未来的城市舞台。天津城市因河而兴、文化中心因水而灵，市民以水为乐。夏季随着音乐喷泉的起舞，每晚聚集到此赏水纳凉的市民有数万人，其活力和吸引力是设计之初未曾完全预见的。智者乐水，仁者乐山，在以水为空间核心的城市设计指导下，简单明确的设计结构，提升了空间场所的亲和力，减少了现代文化集群建设追求极致高冷而带来的排斥感。大剧院的整体屋顶结构统领三个剧场，既满足了使用和运营的需求，又在形式上与自然博物馆形成东西呼应。图书馆"举重若轻"的结构系统将城市空间引入内部，营造出读书乐园的智慧空间；美术馆前庭院的设景，使湖、岸、馆建立了相互关联的空间依赖，宛如天成；博物馆的六重门及内部坡道，将历史、现在和未来展现在游客面前。

本着"贵在至简，精在体宜"的共识，设计团队协调国内外设计责任人对单体高度、间距、材质、共享空间等做出统筹解决方案，保证了集群设计在"和而不同"原则下单体建筑形式的协调和空间上的丰富。

Tianjin Wei(old name for Tianjin) , located in the lower reaches of the nine rivers which flow into the Bohai Sea, has been considered as a paradise for people to realize their dreams. From the time of the Tianzi(royal family) ferry to the establishment of Tianjin Wei more than 600 years ago, it has been known for abundant water and dense marshes. During the Republic of China, people from all over the world gathered in Jingu(anther old name for Tianjin) to create an ideal living environment, such leaving many historical remains as Wudadao, the Italian District, and the Old Town Hall. Within the area the rivers are densely covered, so the city is built beside water and flourished along with water. At the beginning of the 21st Century, Haihe River became the engine of urban development of Tianjin and reconstruction of Haihe River opened the prelude to urban development. With the establishment of the Bohai Sea Economic Circle, the Tianjin Government decided to build the Tianjin Cultural Center to show the development achievements since the economic reform and opening up, and as a demonstration of the city's ambition to be a great cultural city.

Tianjin culture is typically northern Chinese but also mixed with cultures from other countries. By reviewing its modern history, the travails and glory of Tianjin can be fully and truly experienced. The spirit of Tianjin people is one of indomitable optimism in the face of hardship, of modernity mixed with tradition. The regional culture of Tianjin has made it an open, tolerant, and advanced fashion center of the last century. In line with Tianjin's regional cultural genes, the layout of the cultural center is integrated with the axis of the newly built areas and built-up areas. Tianjin's natural

characteristics are reflected by the artificial wetlands, hills and towers, which make the whole site a city square that is both urban ritualistic and reflective of the culture and affinity of the local people.

With the original water surface of the site as the center, the cultural buildings are situated near the water. On the south side, there is a hard and straight shoreline combined with sculptures allocated for museums, art galleries, and libraries. The north side of the slightly curved shoreline is the planned commercial center. On the west side, the artificial wetland grass slopes to extend to the water, and the hills and water combine to create green barriers. The Grand Theatre serves as a background to the east side, creating an artificial water-friendly platform to fulfill people's dream of living near the water, which is also planned to become a future-oriented urban stage. Tianjin is built by the river, the cultural center is built by the water, and the citizens enjoy the water. In summer, thousands of people will gather here every night to enjoy the water and cool off at the music fountain, the vitality and attractiveness of the compound were not imagined at the beginning of the design.

The wise enjoy the waters, the benevolent enjoy the mountains. Under the guidance of urban design with water as the core of space, simple and clear design structure improves the affinity of space places. It reduces the sense of exclusion caused by the pursuit of alienation in the construction of modern cultural clusters.

鸟瞰图 | aerial view

The roof of the Grand Theater is a giant structure covering three theaters, which not only meets the needs of their operation but also matches with the Nature Museum on the west side. The library's 'light' structural system integrates the urban space with the inner space, creating a place for reading. A courtyard is also set in front of the art museum so that the lake and the shore pavilion establish space as naturally formed. The museum's six doors and internal ramps show the history, present, and future to visitors. Based on the consensus that "simplicity is the most important thing, the preciseness lies in the appropriate size". Moreover, we coordinated with those responsible for each design to create a comprehensive solution in regard to the height, spacing, material, and public space of the monolithic building. This ensures the coherence of the monolithic building form and enriches the special feature of the space according to the cluster design's principle of " Being Harmonious yet Different ".

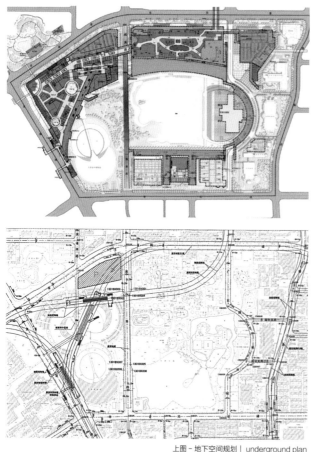

上图 - 地下空间规划 | underground plan
下图 - 轨道交通规划 | rail transport plan

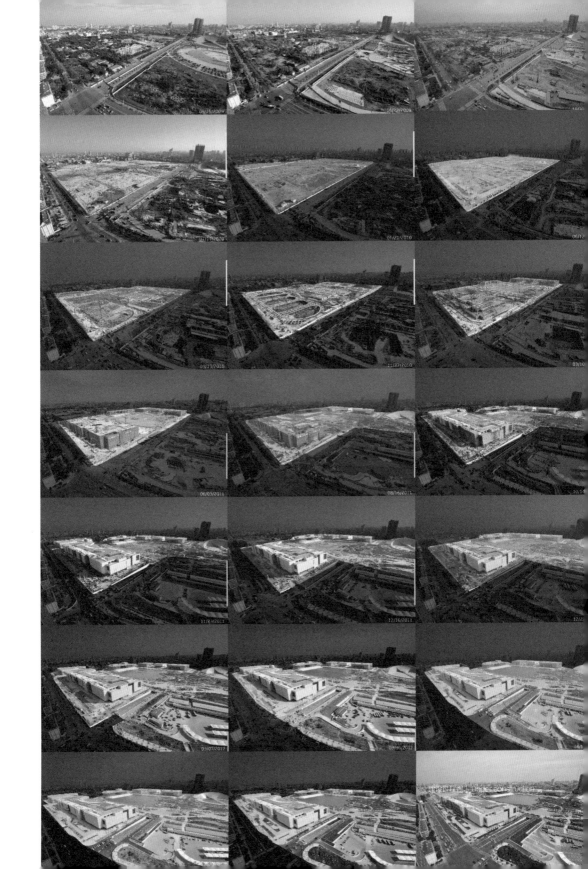

天津大剧院
Tianjin Theatre

入口广场 | entrance square

入口图 | entrance

天津大剧院，总建筑面积 90 000 m²，地上五层，地下一层，高 32 m，由综艺剧场、音乐厅和多功能厅三座独立使用的体块组成，其中综艺剧场内有 1 600 个座位，音乐厅内有 1 200 个座位，多功能厅内有 400 个座位。开阔舒畅的室外临水平台，为人们提供了一个半室外休憩、观景及远眺空间，非常适合搭设临时舞台，可提供灵活的表演场地，是一个绝佳的开放的城市舞台。天津大剧院与自然博物馆位于湖区的两端，自然博物馆为碟状造型，从地面缓缓升起，大剧院如同一片云"悬浮"在大地景观之上，形成了形式上的对话——天与地。

Tianjin Theatre, with a total construction area of 90,000 square meters, has five stories above the ground and one story below. It is 32 meters high and consists of three blocks: a theatre, a concert hall, and a multifunctional hall. The theatre has 1,600 seats. The concert hall has 1,200 seats and the multifunctional hall has 400 Seats. The open outdoor platform provides a semi-outdoor space for sitting and viewing, ideal for setting up temporary stages and providing flexible open stage. The Nature Museum located at the other side of the lake, in the shape of a saucer rising gently from the ground. Tianjin Theatre floats like a cloud above the landscape, symbolizing the dialogue between heaven and earth.

实景图 | scene photo

天津图书馆
Tianjin Library

天津图书馆，建筑面积约 55 000m²，地上五层，地下一层，高 30 m，藏书 600 万册，日读者流量 5 000 ~ 8 000 人次。功能包括文献信息资源收集、整理、存储、学术研究、教育培训等，是承载智慧的建筑，是市民获取知识的乐园。其阶梯状的露台，如同悬浮在空中的书架，组成了内部空间结构。外墙石材百叶有效阻挡阳光的直接射入，消除了 100 m×30 m 巨大立面带给人的压迫感，营造出宁静的氛围，塑造出新颖且透明感十足的石材立面效果。斜坡状的裙房，铺满草坪，为原本方正的建筑物增添了几分温柔恬静。

Tianjin Library has a building area of 55,000 square meters, with five floors above the ground and one underground, and is 30 meters high. It has a collection of 6 million books and 5,000 to 8,000 people visit here per day. Its functions include collection, collation, storage of literature resources, academic research, education, and training, etc. It is a paradise for citizens who love reading. Elements such as the stepped terrace and the bookshelves suspended between the terrace build the internal space structure. The stone louvers effectively cut off sunlight, eliminating the oppressive feeling of the large facade (100 m×30 m) and creating a tranquil atmosphere. The sloping podium, covered with lawn, soften the square building.

外部图 | exterior

中庭 1 | atrium 1

阅读平台 | reading platform

中庭 2 | atrium 2

天津美术馆
Tianjin Art Museum

外部图 | exterior

内部实景 | interior

天津美术馆,建筑面积约28 000m², 地上四层,地下一层,高30 m。设中国书法、西方艺术、雕塑及现代艺术四个长期展厅和多个巡回展厅,可在这里进行学术交流培训,以及美术作品的展示、研究和收藏。现收藏珍贵藏品1万件,每年举办50场高水平展出。在馆前湖岸林荫大道扩展出一个室外艺术展示平台,参观者由此经过入口庭院进入美术馆内部,艺术大厅将室内外空间连成一体,强化人们从自然环境到建筑内部的体验,实现了自然与文化的空间转换。

Tianjin Art Museum has 28,000 square meters, with four floors above ground and one underground, and is 30 meters high. There are four permanent exhibition halls for Chinese calligraphy, Western art, sculpture, and modern art and several traveling exhibition halls for academic exchange training, display, research, and collection of artworks. When completed, the museum plans to collect 10,000 pieces of precious artwork within five years and hold 50 exhibitions of high-quality artwork every year. An outdoor art display platform extends from the lakeside in front of the museum, from which visitors will enter the museum through the entrance courtyard. The art hall connects the indoor and outdoor spaces, showing the process of experiencing from the outside to the inside of the building and reinforcing the spatial transformation of nature and culture.

入口中庭 | entrance atrium

天窗 | skylight

外部图 | exterior

天津博物馆
Tianjin Museum

天津博物馆，建筑面积约 55 000m²，地上五层，地下一层，高 30 m。包括基本陈列、引进展览、学术交流厅、文物库房及办公空间等几个部分，是集文物收藏、保护、展示、研究于一体的综合性博物馆，馆藏达到 20 个门类 20 万件，图书资料 20 余万册。以"六重门"的设计寓意天津设卫建城 600 年的历史。全馆以"世纪之窗"作为空间主题，既是"回顾天津设卫建城 600 年的文明之窗""再现中华百年看天津的历史之窗"，也是"展望天津美好前景的未来之窗"。内部空间用 30 m 宽、14 m 高、100 m 长的时光隧道大厅串联各展厅。由入口通过 93 级台阶直达未来之窗，两侧设展台，令观众在行进中感受时光荏苒，岁月变迁。

Tianjin Museum, covering 55,000 square meters, has five floors above ground and one underground, and is 30-meter high. Providing the permanent display, the introduction of exhibitions, academic exchanges, cultural relics storage, and office and other part. It is a comprehensive museum whose collection reached 20 categories 200,000 pieces, and has more than 200,000 volumes of books. Its six doors signify the 600 years since the founding of Tianjin and the windows that allow you to see the history and future. The internal space is linked to the exhibition halls by a 30-meter wide, 14-meter high, and 100-meter long tunnel. The entrance to the future window is reached by a 93-step staircase, with booths on both sides, allowing visitors to experience the changes in time as they walk.

入口图 | entrance

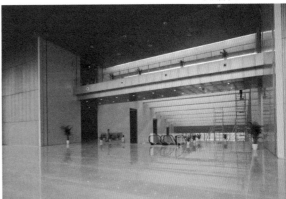

内部实景图 | interior scene

中庭平台 | atrium platform

内部灯光实景图 | interior light scene

入口广场 | entrance square

外部图 | exterior

彩悦城
Sunshine Paradise

彩悦城建筑面积约 82 000 m²，地上五层，地下一层，高 29.55 m。内设室内游乐区、摄影馆、动漫城和儿童仿真体验馆、电影院等，是面向青少年的寓教于乐的综合基地。该建筑呈三叶草造型，功能组团为叶片，公共空间为叶脉。采用夸张的积木阳台着以鲜艳、高饱和度的色彩，配合彩虹般炫丽的地面，营造出童话般的儿童世界。

The sunshine paradise covers 82,000 square meters, with five floors above the ground and one underground, and is 29.55 meters high. It has indoor amusement, animation city, children's experience hall, cinema, and other parts. It is a comprehensive station for adolescent to have fun and learn knowledge. The architectural theme is clover. The functional areas can be compared to the blades, and the public space is the leaf veins. The exaggerated block balconies and bright, pure colors, together with rainbow-like floors create a fairy-tale children's world.

中庭 | atrium

室内中庭 | atrium

外部图 | exterior

万象城
Tianjin MIXC

万象城建筑面积 370 000 m², 地上五层, 地下两层, 高 33 m, 拥有以 "日辉、月光、星光、大地"为主题的四个共享中庭, 整体效果亲切、宏大, 是集购物、餐饮、娱乐、休闲、文化等功能业态于一体的国际化购物中心。

With a construction area of 370,000 square meters, Tianjin MIXC has five floors, 33 meters high above ground, and two floors below ground. It has four intimate and grand atriums. The themes of the atriums are "Sunlight, Moonlight, Starlight, and Earth". It is an international shopping center that combines shopping, dining, entertainment, and other multi-functional businesses.

光伏发电系统位于万象城的屋顶，采用支架平铺7 400 m² 非晶硅薄膜电池，额定发电功率 300 kW，系统输出分 15 路，分别并入变电站内照明配电柜的低压母线上，为地下车库照明提供绿色能源。

利用中水和雨水建设中心湖，面积 100 000 m²，平均深度 1.7 m 在市民熟悉的空间场所的内部，集合了能源站、地热资源的地源热泵、湖水的生物净化系统、地下雨水收集系统等绿色节能技术的应用设施，使文化中心的内涵从单一文化服务上升到文化与技术交融、人与自然共生的层面，体现了新时代津派文化开放、包容、共享的精神实质。

The Solar power generation system is located on the roof of the MIXC, with 7,400 square meters amorphous silicon thin-film cells laid on a bracket, rated power of 300 kW. The output is divided into 15 channels, which are integrated into the low-voltage lines of the lighting distribution cabinet in the substation to provide energy for the underground garage lighting.

The central lake is constructed by wastewater and rainwater, with an area of 100,000 square meters, an average depth of 1.7m. There are green and energy-saving technologies, such as the integrated energy station, ground-source heat pump for geothermal energy, a biological purification system for lake water, and underground rainwater collection within the space place known by the public. These technologies ensure the cultural center is not only a cultural service place integrating culture and technology, but also a demonstration of a symbiosis between human and nature, reflecting the spirit of new era Tianjin culture: openness, tolerance, and sharing.

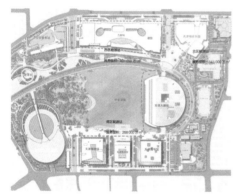

能源站分布 | energy stations

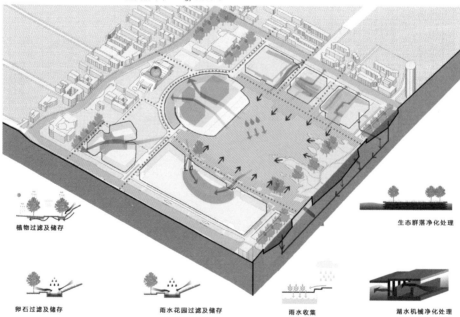

湖水循环净化处理系统 | lake water recycling purification treatment system

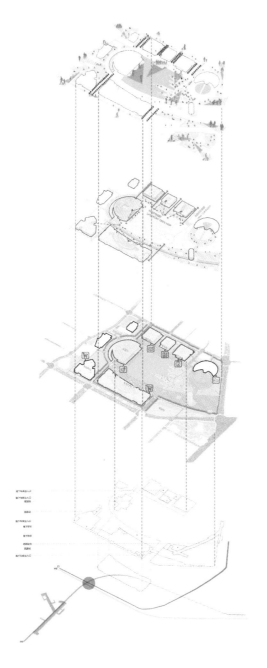

绿地渗透
Green Penetration
开放的公共文化空间鼓励和引导公众进入。文化设施的布局与城市绿地的整合联动,提升了该区域的城市活力。

人行系统
Pedestrian System
岸线、绿道、广场、林荫,划分出一系列户外活动场所。弧形轴线勾勒出中央湖面北侧的城市步道与市民广场岸线。南岸文化公园、西岸自然生态公园与东岸城市舞台亲水空间互为因借。

空间布局
Space Layout
文化中心空间布局以湖为核心,以水体、绿植为主进行园林式布局。

地下连通
Underground Connection
地下空间作为整体进行开发。将地下空间包括市政道路下方进行了全开发,实现地下车库互通互联,有效减少地上车库出入口的设置。综合利用地下空间解决了设备用房和商业配套的停车需求,同时也提供了地铁站接驳功能和其他商业功能。

爆炸图 | exploded diagram

滨海新区建设是国家继深圳、浦东开发之后的又一战略重点。伴随滨海新区内北方国际航运中心的建成,于家堡商务中心的落成和北京直达滨海新区高铁的投入使用,作为配套基础设施的文化场馆建设成为日益紧迫的任务。初期选址的变化使得第一次国际方案征集的成果只能作为参考,同时天津文化中心的落成成为当时滨海文化中心对标的方向。滨海新区的经济、人口、环境等因素成为决定文化中心布局模式的先决条件。天津文化中心依托直辖市在布局上注重礼仪性,宏大叙事的方法是否适合滨海新区文化中心的建设成为最先思考的问题。

当时在经济总量上,滨海新区的强劲实力体现在人均 GDP 方面远高于中心城区,但滨海新区在人口规模和密度、环境承载力、交通配套设施等方面与中心城区差距明显,这些客观条件让规划者面对现实冷静思考之后,采取收缩规模容量的策略应对现实的挑战。滨海新区选择与天津文化中心建设城市广场不同的方式,通过集合文化建筑增加商业氛围,营造富有活力的城市客厅的方法,探索文商融合建设的新模式。

天津滨海文化中心
Tianjin Binhai Cultural Center

用地面积:12 hm²
建筑规模:3 120 000 m²

城市设计团队:
天津市城市规划设计研究总院

修建性详细规划团队:
天津市城市规划设计研究总院

获奖情况:
2019 年度天津市优秀城乡规划设计奖
2017 年度全国优秀城乡规划设计奖二等奖
2017 年度天津市优秀城乡规划设计类(城市规划类)二等奖
2017 年度全国优秀城乡规划设计奖(城市规划类)表扬奖

鸟瞰图 | aerial view

区位图 | location

功能复合的阻力来自传统文化事业与产业对峙的思维模式,长廊的设置客观上为功能融合提供了物理空间,同时在沿长廊的文化建筑的首层设置商业空间,保证了长廊的界面活力,这一点在许多商业街道的设计中已被广泛证明。文化中心建成之后,长廊空间因其承载了不断从原有功能上拓展出来的文化行为以及新兴的文化活动,成为滨海新区最具活力、弹性、魅力的存在。长廊实现了文化事业与文化产业的共生共荣,不但从使用上给文化空间带来大量人气,同时商业空间的开发给文化中心的运行也带来了持续的经济保障和社会公益价值。

滨海文化中心的建成解决了滨海新区市民文化消费场所匮乏的问题,它从开放至今一直被广泛关注并成为展示滨海新区创新发展的决心和成果的重要场所。回顾其建设历程让我们对那时的决策与建设者们的勇气与担当钦佩不已。创新是这个时代的主旋律,滨海文化中心在空间模式、功能模式及绿色模式方面还有更多值得总结的地方。

The construction of the Binhai New Area was the strategic focus on the development of China after Shenzhen and Pudong. Along with the completion of the international shipping center of Binhai new area, the implementation of the Yujiapu Business Center, and the opening of the high-speed railway station from Beijing to Binhai, the construction of cultural venues as supporting infrastructure became an increasingly urgent task. The change of the initial site selection made the first international call for proposals only a reference. Besides, the completion of the Tianjin Cultural Center became the guidance of how the Binhai Cultural Center should be built at that time. The Binhai New Area's economy, population, and the environment became prerequisites for deciding the layout of the cultural center. Because the direct municipality under the Central Government, the Tianjin Cultural Center focuses more on rituals and a grand narrative approach. Whether this can be the model of Binhai Cultural Center became the first question to think about.

At that time, the per capital GDP of Binhai was much higher than that of the central city. However, there is a gap between the Binhai New Area and the central city in terms of population scale, density, environmental carrying capacity, and transportation support. These conditions made the planners contract the limited letter of the issue to cope with the challenges after facing the fact and thinking calmly. By choosing a different approach from the construction of the city square in the Tianjin Cultural Center, a new model of cultural and business integration is explored through the collection of cultural buildings to build the commercial atmosphere and create a vibrant city living room.

By the principle of pragmatism and efficiency, a new version of the "Five Pavilions and One Corridor" urban design was formed under the guidance of the Planning Bureau, relying on the core area of Ziyun Park, as a guide for the second

international call for proposals. Unfortunately, due to the slot, Zaha Hadid, who participated in the first call and won the bid, died suddenly and did not leave her work in Tianjin Binhai. The new cultural center is based on the model of a cultural promenade including various cultural buildings to build a cultural street. Developing commerce on the cultural basis, the cultural venue with commercial space, and the energy center with shared resources would afford more commercial function and a model of environmentally friendly multi-line transportation would be built.

区位分析图 | site analysis

场地周边功能 | function around site

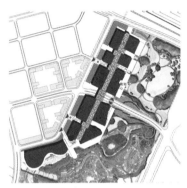
功能整体布局 | general function arrangement

夜景鸟瞰图 | aerial view in night

The resistance to functional compounding comes from the traditional mindset of confrontation between cultural undertakings and industries. The promenade objectively provides opportunities for integration in physical space. The setting of commercial space on the first floor of the cultural buildings will increase the interface vitality of the promenade, which has been tested in the design of commercial streets. After the completion of the Binhai cultural center, the promenade space becomes the most dynamic, resilient, and charming presence because it carries original and emerging cultural activities. The promenade realizes the symbiosis and co-prosperity of undertakings and cultural industries. It makes more people go to the cultural space. Besides, it brings continuous economic income and social welfare value to the operation of the cultural center.

The Binhai Cultural Center solves the problem of weak cultural consumption of the citizens of the New Area, and it has been widely watched since its opening and has become an important place to show the determination and achievements of the innovation development of Binhai New Area. The construction process has made us admire the courage and commitment of the decision-makers and builders in the current environment. Innovation is the dominant theme of this era, and the Binhai Cultural Center spatial model, functional model, and green model have more to be summarized.

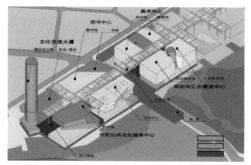
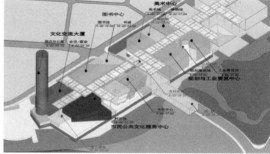

功能分析图 | function analysis

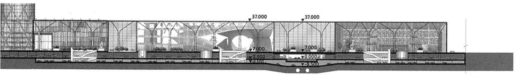

剖面图 1 | section 1

剖面图 2 | section 2

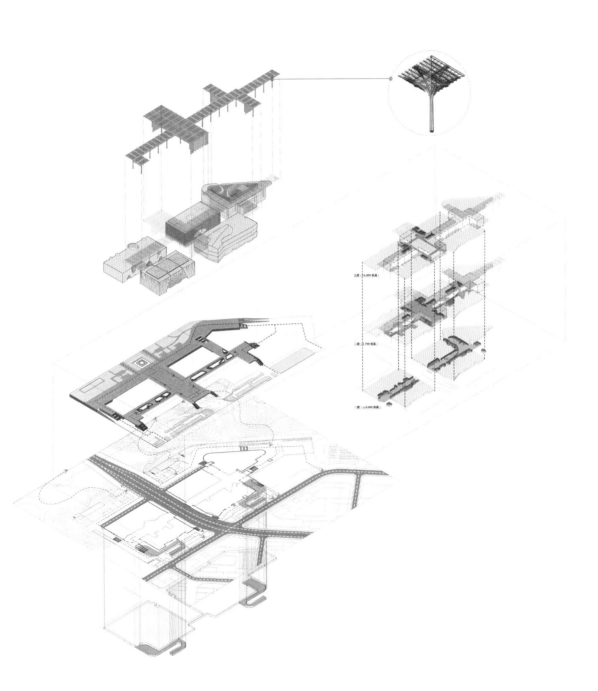

爆炸图 | exploded diagram

滨海图书馆
Binhai Library

滨海图书馆，建筑面积约为 33 700 m²，地上 5 层，地下 1 层。基于对新时代图书馆模式的理解，设计师在公共空间设计中提出了中庭的设计概念"书山"。"书山"不仅联系各个功能区块，同时营造了承载交流、休憩、观览等活动的公共空间。另一个设计概念"滨海之眼"则更多地考虑了室内与室外、建筑与城市之间的关系。直径 21 m 的球体报告厅，以人体尺度为基本参数，"书山"曲线围塑知识海洋的共享空间，为中庭带来了强烈的视觉冲击。整个共享空间使长廊与公园融为一体，进而成为滨海新区的城市名片。

Binhai Library, covering 33,700 square meters, has 5 floors above ground and 1 floor underground. Based on the understanding of the new era library model, the designer proposed the atrium design concept of "Book Mountain" in the public space design. The "Book Mountain" not only connects various functional areas but also creates a space for communication, resting, and viewing activities. Another design concept, "The Eye", takes into account the relationship between indoor and outdoor, and between the building and the city. The 21-meter diameter spherical lecture hall takes the human scale as the basic parameter and fits the curve of book mountain to shape the shared space of the knowledge ocean. It creates a strong visual image for the atrium and makes the promenade and the park integrated, thus becoming the city card of Binhai New Area.

立面图 | facade

"大眼睛"中庭 | "eye" atrium

中庭阶梯 | atrium ladder

滨海科技馆,建筑面积为 32 730 m², 地上 4 层,地下 1 层。滨海科技馆的选址是在天津碱厂旧址的南面,从 20 世纪 20 年代开始,两个圆锥形混凝土冷却塔便矗立在天津碱厂之中。它见证着碱水的流出和工人们汗水的挥洒,成为岁月和历史的象征。为了向历史致敬,设计师们基于场地悠久的工业历史构思滨海科技馆的设计,抽取了"锥形圆筒"的体量符号,并将其作为方案概念的起源,运用现代的建筑材料和手法再现了老碱厂的光辉岁月。科技馆以科技成果展示和科技知识普及教育为主要功能。

Binhai Science and Technology Museum, covering 32,730 square meters, has 4 floors above ground and 1 floor underground. It is in the south of the former site of the Tianjin Alkali factory. Since the 1920s, two concrete cooling towers have been in the middle of the Tianjin Alkali factory. The cooling towers have become a monument of the Tianjin Alkali factory for years. To commemorate the history, the designers took the conical element as the origin of the program concept and used modern construction materials and techniques to recreate the glory of the old alkali factory. It takes the display of scientific and technological achievements and the popularization and education of scientific and technological knowledge as its main functions.

滨海科技馆
Binhai Science and Technology Museum

中央圆筒 | central cylinder

公园侧立面 | facade along the park

流动展厅 | flowing showroom

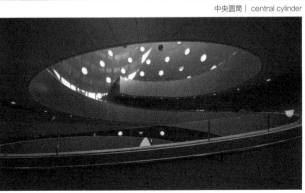

圆筒坡道 | cylinder ramp

长廊入口 | promenade entrance

滨海美术馆
Binhai Art Museum

滨海美术馆，建筑面积为 26 990 m²，地上 4 层，地下 1 层。滨海美术馆以"水晶体"的艺术形态为设计创作要素，整体犹如一座精美的雕塑作品，以精准的比例"切割"出立面的层次变幻，形成独特的虚实交织的立方体，极具魅力。这里展示了当代国内外优秀美术作品，突出了塘沽、汉沽和大港版画传统艺术所蕴含的地域文化。美术馆还通过引入市场化"艺品藏拍"的文化交流产业，以组建艺术家沙龙等形式，提升了功能性与开放度，成为文创交流、艺术育人的重要场所。

Binhai Art Museum, has 26,990 square meters, four floors above ground and one floor underground. It takes "crystal" as the design element, and the whole building is like an exquisite sculpture work. The facade is cut into various layers, forming a unique charming cube. It exhibits not only domestic and abroad modern art, but also the traditional art of Tang'gu, Han'gu, and Da'gang. The Binhai Art Museum also enhances functionality and openness by introducing market-oriented industries and forming artists' salon studios, becoming an important place for cultural and creative exchanges and art education.

展厅 | exhibition hall

外立面 | facade

滨海演艺中心
Binhai Grand Theatre

滨海演艺中心，建筑面积约为 23 980 m²。设计师创作了以"船"为设计主题的建筑形态。出挑的屋顶结合面对城市广场的圆角体形犹如一艘引领整体文化中心的"文化航母"。建筑表面"鳞片"状的幕墙单元造型犹如水面泛起的"涟漪"，更为这艘"文化航母"增添了航行的动感。滨海演艺中心内设有专业演出、群众文化娱乐及艺术创作教育等多方位的综合性文化演艺设施。其内功能包括一座 1 200 个座位的专业歌剧院，一座 400 个座位的多功能实验剧场以及一所小型的音乐、表演和舞蹈艺术创作学校。

Binhai Grand Theatre has 23,980 square meters. The designer created the shape form with the theme of "ship". The prominent roof is like a "cultural aircraft carrier" leading the overall cultural center. The "scaled" curtain wall units on the surface of the building are like "ripples" on the water surface, adding a sense of motion to this "cultural carrier". The Binhai Grand Theatre is a comprehensive cultural and performing arts facility covering professional performances, mass cultural entertainment, and artistic creation and education. It includes a 1,200-seat professional opera house, a 400-seat multi-functional experimental theatre, and a small school of music, performance, and dance art creation.

内部剧场 | internal theatre

外立面 | facade

中庭阶梯 | atrium ladder

滨海市民中心
Binhai Civic Center

滨海市民中心，建筑面积为 42 300 m^2。设计团队在不同楼层设计了不同的倾斜角度，对立面构件也进行了精心的排列设计，以此来呼应"水波浪"的主题。室内通过一个阶梯形共享空间创造出一个室内的四季花园，打造"滨海绿洲"的主题。滨海市民中心是一个微缩的"文化中心"，剥离原"文化馆"定义中与专业的文化场馆相重叠的功能单元，保留其他专业场馆无法提供的功能，重新定义教育培训的方式，由被动式的授予转化为参与互动和开放创新的自由模式。立体斜向共享空间将各种功能整合起来，打破了办公空间呆板的形象，展现了绿色生态活力空间的魅力。

Binhai Civic Center covers an area of 42,300 square meters. Folded corners are designed on different floors and the facade is carefully arranged to show the idea of "water waves". The garden in the shared space shows the theme of the "coastal oasis". Binhai Civic Center is a "cultural center" that removes the original functional units. It retains the functions that other professional stadiums do not have, and redefines education and training, transforming from passive teaching to interactive and open mode. The three dimensional shared space integrates various functions, breaking through the dull image of office space and showing the charm of green ecological vitality space.

景观中庭 | atrium

外部入口 | external entrance

内部入口 | interior entrance

文化长廊
Cultural Corridor

文化长廊的创新设计将传统意义上的文化中心转化为以长廊为核心的文化建筑集群,各个文化建筑单独存在的意义被不断弱化消解,同时,彰显了长廊在文化街道的价值,与各个单体的公共空间共同构成了城市客厅,成为文化集群共享空间的核心,承载着新时代文化活动的需求。在推敲尺度的过程中,设计团队以街道设计的经验优化步行的连续性和体验的趣味性,以人的行为作为依据,增强空间开放度、节点合理性(在长廊宽度为18 m、24 m、30 m的方案比选中均有体现)。

The establishment of the cultural corridor transforms the traditional cultural center into a cluster of cultural buildings with the promenade as the core. The importance of each cultural building alone is weakened and gradually grows into various cultural activity areas. Meanwhile, the culture corridor highlights the value of the cultural street. Together with the public space of every single building, it constitutes the city living room and becomes the core of the cultural cluster. During scale designing, the experience of street design is used by the design team to optimize the interest of the walking experience. Also, human behavior is fully taken as the basis for the design and nodes (promenade widths of 18, 24, and 30 meters are proposed).

文化长廊 | cultural promenade

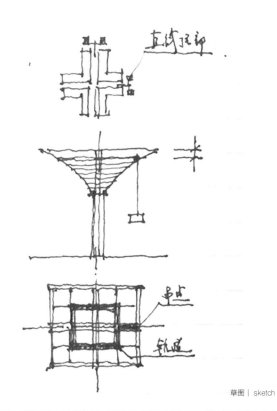

草图 | sketch

结构细部 | structure detail

幕墙效果 | curtain wall

文化长廊可持续设计目标
Sustainable Design

最大化利用自然光，保持长廊内适宜的风环境，冬季利用被动式太阳房设施辅助采暖。

Maximize natural lighting, keep pleasant wind environment in the corridor, use passive solar house facilities to assist in winter heating.

文化长廊风环境分析
Analysis of Wind

采用可灵活调节的设计方案：通风口冬季能够完全封闭，夏季和春秋季能够打开。

Flexible and adjustable design scheme is adopted: the ventilation opening can be completely closed in winter, while it can be opened in the other seasons.

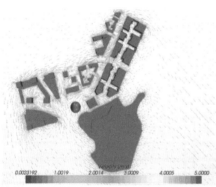

冬季全封闭分析 | winter fulled lock analysis

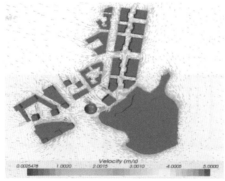

夏季全封闭分析 | summer fulled lock analysis

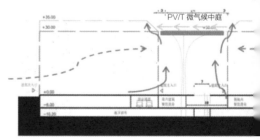

夏季-PV/T 微气候中庭 | summer-PV/T microclimate atrium

文化长廊 | cultural corridor

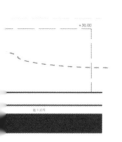
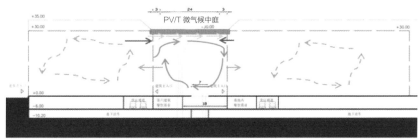

冬季-PV/T 微气候中庭 | winter-PV/T microclimate atriumr

. 济宁位于山东省西南部,有孔庙之乡、运河之都的美称,是鲁西南中心城市。我们有机会参加济宁文化中心建设始于一次偶然的学术交流。那是 2012 年天津文化中心建成之后举办的一次文化中心设计论坛,主办方邀请各方代表参会交流,时任济宁主管建设的领导莅临会场,交流中对我们全过程咨询设计产生兴趣,同时对建设和管理也特别关注。随后受他们邀请参加济宁文化中心设计投标,当时正值我们参加滨海文化中心设计,虽然时间上有冲突,但我们被他们的诚心深深打动,就此我们与济宁结下不解之缘。

济宁市有深厚的文化底蕴,近年来经济发展取得很大成就。文化中心选址开发区风景迤逦的太白湖畔,虽然自然资源禀赋丰厚,但人口密度低,交通基础设施不完善,场馆的规模有限,建设资金也有限,这成为设计的难点也是出发点。团队集思广益总结归纳,提出"文商旅融合"的功能模式,结合"城市文化高地"的空间构想,为济宁量身定制了面向未来的城市文化综合体。

济宁文化中心
Jining Cultural Center

用地面积：53.5 hm^2
建筑规模：491 000 m^2

城市设计团队：
天津市城市规划设计及研究总院建筑一院
德阁（RSAA）建筑设计咨询有限公司
美国 LLA 建筑设计公司

修建性详细规划团队：
天津市城市规划设计研究总院建筑一院
济宁市规划设计研究院
华东建筑设计研究总院

获奖情况：
2019 年度天津市优秀城乡规划设计奖（城市设计类）三等奖
2016 年度创新杯建筑信息模型设计大赛绿色设计优秀 BIM 应用奖
第十三届中国钢结构金奖

鸟瞰图 | aerial view

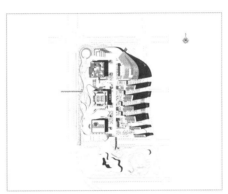

区位图 | location

"孝人敬天"的古训在规划上得以传承，但古代权威礼序在这里转化为生态礼序，除文化建筑的间距、院落关系与视线通达以外，现场风环境的分析也影响了建筑的具体位置。同时，自然的融入也是优先考虑的因素，朝向太白湖，每个单体都要有面向湖面的共享空间，还要有观景空间，从平台、出入口、外廊空间，到内部共享、独立空间，都以生态礼序为先。

空间上通过设置的平台，实现"双首层"面向自然开放，平台上文化与商业相连，平台下辅助功能相连互补。西"文"、东"商"的格局使文化中心朝向太白湖一侧展现文化建筑的风雅，邻接城市干道的一侧展示出商业建筑的活力。将"文""商"建筑最大限度地融合，两者互为依托展现济宁文化悠远的过去、商业繁华的未来。

Jining is in the southwest of Shandong Province, known as the river's city, the hometown of Confucius Temple, and the center of southwest Shandong. The opportunity to participate in the construction of the Jining Cultural Center began with an academic design forum in 2012, after the completion of the Tianjin Cultural Center. During the forum, the leaders of Jining were interested in our design and the construction and management. Then we were invited to participate in the design bidding of Jining Cultural Center. At that time we were designing the Binhai Cultural Center, so there was a contradiction in time. But we were impressed by their sincerity, and finally, we set an inseparable relationship with Jining.

Jining has great cultural heritage, and the economic progress of Jining was good in recent years. The cultural center is in the development zone on the shore of Tai Bai Lake, where the natural resources are rich. However, the population density and transportation infrastructure are insufficient. The inadequate scale of the venue and controllable operation cost became the difficulty of the design and the starting point. The "cultural and commercial separation model" of Tianjin Cultural Center and the "cultural and commercial combination model" of Binhai Cultural Center can no longer solve the problems when planning Jining Cultural Center. In response to that, the team pooled their wisdom and summed up the problems. They proposed the "cultural, commercial, travel mixed model". Put forward the functional

model of "integration of culture, business, and tourism" by adding up the concept of "urban cultural highland", the team customized the future-oriented urban cultural complex for Jining.

In this design, the tradition of filial and piety was reflected in the respect people have for the nature. In addition to the building spacing, courtyard relationships, and sightline access, wind also influenced the design. At the same time, the overall environment was also considered. Each building has spaces facing Lake Taibai, such as terraces, entrances and exits, exterior corridor spaces, shared interior, and independent spaces. These spaces are arranged according to the traditional "etiquette" concept while assuming ecological functions.

By setting the platform, the "double first floor" is open to the public. The commercial and culture are above the platform, while other functions are under the platform. The commercial part locates in the east and the cultural part in the west. It makes the one side next to Taibai Lake more elegant and the other side next to the main road more energetic. The maximized integration of cultural and commercial buildings shows the future of Jining's long-standing culture and commercial prosperity.

aerial view

规划设计方案演变
Development of Master Plan

设计团队对单体布置在间距、尺寸、材质、色彩、形式上进行统筹思考,将充满张力的大屋顶的图书馆放在中间,以从空间上统领文化集群,同时图书馆底部收紧,以从空间上缓解压力。临平台一侧文化建筑地下层,全部设置为商业空间,增加商业机会和活力,平台下设置能源站集中供暖、制冷,地下贮水践行海绵城市理念。

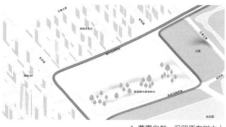

1. 尊重自然:保留原有树木 |
respect nature: preserve the original trees

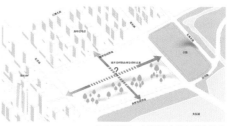

2. 空间过渡:自然空间向城市空间过渡 |
spatial transition: transition from natural space to urban space

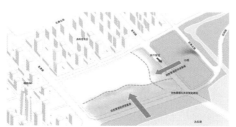

3. 缝合空间:自然推进基地 |
suture space: natural propulsion base

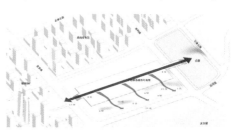

4. 平台联接:平台联系城市与自然 |
platform connection: the platform connects the city and nature

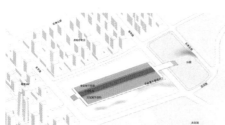

5. 功能划分:文商功能结合 |
function division: combination of cultural display and commercial sales functions

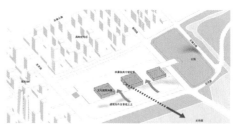

6. 抬升体量:体量抬高方便观景 |
raise the volume: raise the volume to facilitate viewing

7. 功能互促:文商功能互促 |
mutual promotion of functions: cultural display and commercial sales functions promote each other

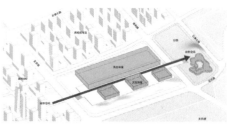

8. 空间互通:城市与自然空间互通 |
spatial connectivity: urban and natural space connectivity

The arrangement of the monoliths is in terms of spacing, size, material, color, and form. The large roof of the library is in the middle to lead the cultural cluster, while the bottom is tightened to provide a larger outside space. Shops and commercial space are on the ground floor of the cultural building facing the platform to increase business opportunities and vitality. Energy stations are set under the platform for heating and cooling. The underground water is stored to practice the concept of "Sponge City".

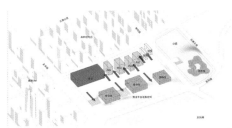

9. 观景限定：限定平台观景空间 |
viewing limit: limit the viewing space of the platform

10. 地下空间：停车与文商功能结合 |
underground space: combination of parking and cultural display and commercial sales functions

11. 置入庭院：广场置入下沉庭院以丰富界面 |
place courtyard: a sunken courtyard is placed in the square to enrich the spatial levels

12. 介入人流：东侧人流进入商业 |
intervention visitors: the flow of people on the east side enters the commercial area

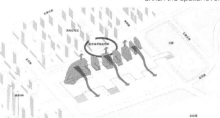

13. 界面延续：街接东西两侧不同空间形态形成界面延续 |
continuation interface: connecting different spatial forms on the east and west sides to form interface continuation

14. 视线交流：建筑界面与景观发生视线交流 |
communication of sight: sight communication between architectural interface and landscape

15. 形态相适：建筑入口形态与环境结合 |
appropriate form: the combination of building entrance form and environment

16. 交通可达：周边道路街接进入场地 |
traffic accessibility: the surrounding roads connect to the site

功能整合
Function Integration

由于济宁的地下水位较高，我们将原本置于地下的地库，上浮到水位线以上，形成高地公园。这样高地之上，建筑与建筑之间形成互动界面，高地之下又形成了商业空间、交通空间、停车空间以及辅助空间。

Due to the high underground water level in Jining area, we have raised the basement, which was originally placed underground, above the water level to form a high platform park. We make an interactive interface between the buildings, with commercial space, transportation space, parking space, and other spaces under the high platform.

功能空间分布 | functional space layout

地上空间整合 | aboveground space integration

地下空间整合 | underground space integration

济宁文化中心长卷展开图 | long scroll drawing

动线整合
Streamline Integration

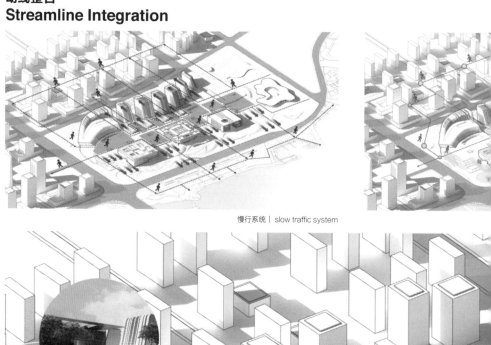

慢行系统 | slow traffic system

公共空间路径 | public space path

生态的文化公共空间 | ecological cultural public space

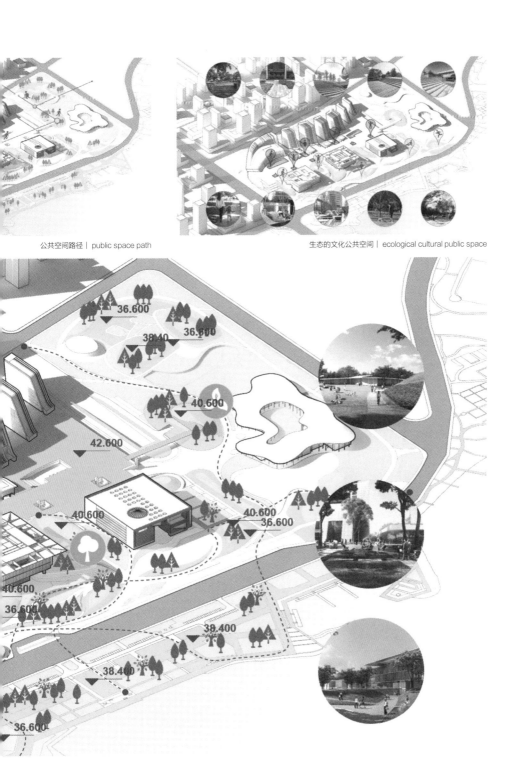

联系自然与城市的有趣路径 | path connecting nature and city

57

功能渗透
Function Penetration

在不同的建筑之间,我们协调各个场馆的经营性与非经营性场所,使得场馆可向社会提供场地出租的服务。

无论在高地公园还是在西侧的文化公园,高低错落的景致形成了曲径通幽的效果,从而形成了联系自然与城市的有趣路径。同时打破建筑之间街墙一样的壁垒,形成东西南北向贯通的多条连续景观通廊,形成畅达连续的步行网络,就这样用"延续"和"打破"来构造文化的、自然的、生态的公共空间。

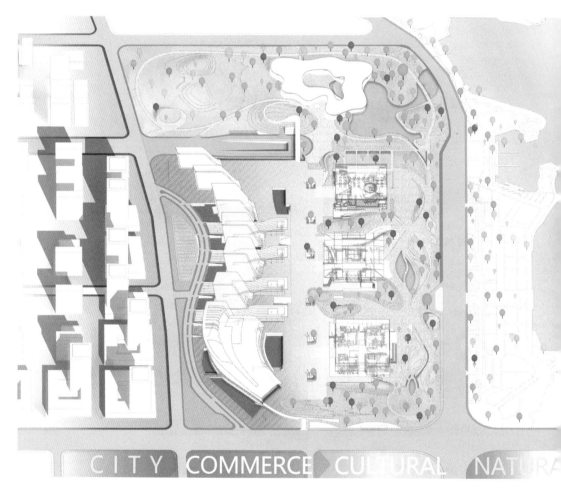

功能统筹分析 | function integration

We counted the number of operating and non-operating sites of each venue to meet the demand for venue rental use.

Whether above highland park or between cultural parks on the west side, the staggered view creates a curvy effect, creating an interesting path linking nature with the city. At the same time break the building between the street wall-like barriers, the formation of east-west north-south through the N-bar continuous landscape corridor, to ensure smooth access to a continuous pedestrian network. In this way with a "continuation" and a "break" formed a cultural, natural, ecological cultural public space.

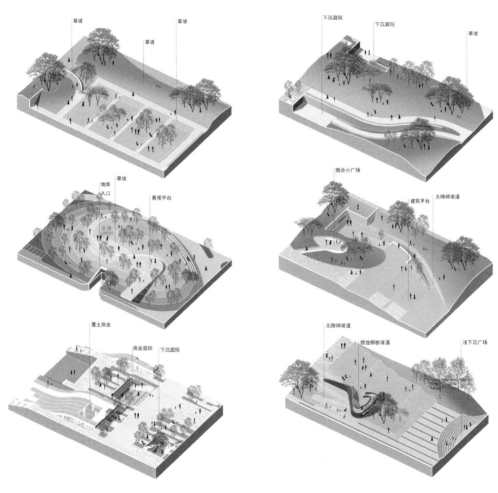

高地公园节点图 | highland park node graph

整体建成效果
Overall Construction Effect

绿谷景观 | green valley

景观鸟瞰图 | landscape aerial view

群众艺术馆一侧

Group Arts Center

图书馆—侧景观 | side view of the library

鸟瞰图 | aerial view

总平面图 | general plan

济宁市群众艺术馆
Jining Group Arts Center

济宁市群众艺术馆,建筑面积约 32 163 m² 是一个综合了展览、娱乐、培训、观演功能的文化活动场所。我们试图通过孔子"游于艺"的理念(《论语·述而》中子曰:"志于道,据于德,依于仁,游于艺。"),建立起某种时空上的穿越,让后人能体会出先贤曾经到达的境界。

Jining Group Arts Center is a cultural activity venue that integrates exhibition, entertainment, training, and performance functions. We tried to establish a kind of time and space travel through Confucius' concept of "seek relaxation and enjoywent in the Six Arts" ("Set your heart on the truth, hold to virtue, lean upon human-heartedness and seek relaxation and enjoyment in the Six Arts " in *Analects of Confucius*). People can appreciate the realm that the sages have reached.

外部图 | exterior

内部图 | interior

中庭 | atrium

济宁市图书馆
Jining Library

济宁市图书馆建筑面积约29 316 m², 以"学宫—辟雍"为设计原型, 打造以杏坛为中心, 自下而上, 由动到静, 逐渐打开的十字对称中庭, 并结合高台明堂、环形水系和园林再现辟雍形式, 全面彰显孔孟之乡济宁的城市文化魅力和作为知识圣地的图书馆的文化职能。馆内设计藏书量约150万册, 将最大限度满足群众阅读需求, 让知识与文化视野在这里升华。

With an area of 29,316 square meters, the Jining Cultural Center library is designed based on the prototype of "Confucius temple and Imperial academy", creating a cross symmetrical atrium with the Xingtan as the center. The library is also combined with a high platform hall, a circular water system, and a garden to reproduce the form of Imperial academy. The library is designed with a collection of about 1.5 million books, which will meet the reading needs of the public to the greatest extent.

图书馆外部 | library interior

图书馆阅览区 | libaray reading area

图书馆阅览区 | libaray reading area

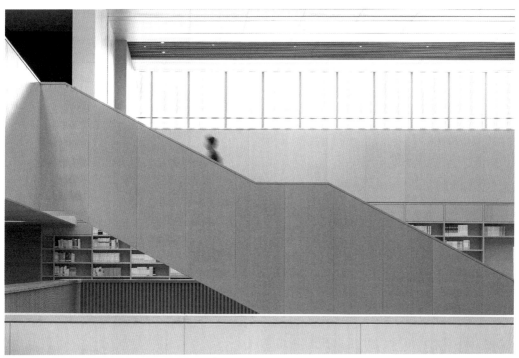

图书馆中庭 | library atrium

图书馆阅览区 | libaray reading area

济宁市博物馆
Jining Museum

济宁市博物馆建筑面积 27 237 m²，设计理念兼顾时代性与济宁地域文化特色，强调空间的流动性与有机渗透，最大限度地将阳光引入室内，让公众充分享受自然与文化的气息，这是一座集搜集、保存、修护、研究、展览、教育、娱乐等功能于一体的综合性、现代化场馆。让人切身感受到济宁多彩的文化魅力和古今沧桑的巨变。

The museum covers an area of 27,237 square meters. The design concept enhances the modernity and regional cultural characteristics of Jining. The fluidity of the space is also important, so we let the sunlight in to allowing the public to enjoy the natural atmosphere. It is a comprehensive and modernized museum integrating collection, conservation, restoration, research, exhibition, education, and entertainment. It allows people to experience the culture of Jining and the changes of the past.

博物馆走廊 | museum corridor

博物馆屋顶 | museum roof

博物馆外部图 | museum exterior

博物馆实景图 | museum scene

济宁市美术馆
Jining Art Museum

济宁市美术馆建筑面积约 8 180 m^2，屋顶外形类似荷叶，与室内回廊空间相协调，室外规划公园广场、庭院水系等，致力打造景观化、休闲化的活力艺术空间。

The roof of Jining Art Museum resembles a lotus leaf, which is in harmony with the indoor space, the outdoor park square, and the courtyard water system. It is set up to create an attractive art space with a landscape.

美术馆鸟瞰图 | aerial view of Jining Art Museum

外部庭院 | courtyard

外廊 | veranda

内廊 | inner corridor

高地公园与绿谷
Highland Park and Green Valley

高地公园位于场馆建筑东侧,是一个高 6 m,南北长约 690 m,东西宽 40 ~ 100 m 的超大平台,下部为停车库及配套服务建筑,它将单体建筑和商业建筑群有机地结合起来,可以说高地公园是建筑之间的一个"共情"空间。

Highland Park is on the east side of the building, where there is a large platform of 6 meters in height and 690 meters in length from north to south. The lower part of the park is the garage and supporting service buildings. It combines the monolithic building and the commercial system. It is a communication space between the buildings.

景观顶视图 | aerial view of the landscape

入口台阶 | entrance step

高地公园 | highland park

景观实景图 | landscape scene

建设过程图
Construction Process Diagram

施工 | construction

施工 | construction

施工 | construction

施工 | construction

施工 | construction

2018年12月1日起实施的国家标准《城市居住区规划设计标准》（以下简称《标准》），调整了居住区分级控制的方式与规模，统筹、整合、细化了居住区用地与建筑相关控制指标，优化了配套设施和公共绿地的控制指标和设置规定。具体解读为以使用者行为作为标准指导居住区规划，交通上推行"窄路密网"，功能上推行"混合住区"，空间上推行"开放社区"，将困扰规划师的人车混流、空间封闭从制度规则上给予化解、引导规划向以人为本的方向发展，为科学规划提供基本的制度保障。河西八大里项目就是在新标准出台之前进行的规划，有幸率先成为天津试行《标准》的实践地。

河西八大里规划是2013年由政府主导的地域开发项目，在当时条件下采用"指挥部领导+设计师集群"的

河西八大里 Badali Community, Hexi District

用地面积：214 hm²
建筑规模：3 730 000 m²

城市设计团队：
天津市城市规划设计研究总院规划八院（愿景城市开发与设计策划有限公司）
天津市城市规划设计研究总院建筑设计一院
天津市筑土建筑设计有限公司
天津市建筑设计研究院
天津华汇工程建筑设计有限公司
天津大地天方建筑设计有限公司
天津市博风建筑工程设计有限公司

修建性详细规划团队：
天津市城市规划设计研究总院
天津市建筑设计研究院

获奖情况：
2017年度天津市优秀城乡规划设计类（修建性详细规划）一等奖
2015年度全国优秀城乡规划设计奖（城市规划类）二等奖
2015年度天津市优秀城乡规划设计奖（城市规划类）一等奖

鸟瞰图 | aerial view

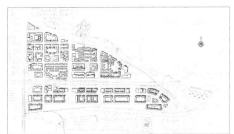

区位图 | location

设计模式。总体城市设计由黄晶涛带领天津愿景城市开发与设计策划有限公司完成。项目用地西起解放路、南至复兴河、北至大沽北路，呈三角形，东西长约2.4 km，黑牛城道从东西向穿过，将用地分为南北两区

规划之初，指挥部提出向国外优秀城市规划学习借鉴先进的规划理念，创新规划方法，如解决现行居住区地块占地过大、各自相对独立封闭、生活配套虽符合标准但不方便使用、交通拥堵、地面人车混行、绿化面积较小且分散、职住严重分离等问题。指挥部组织设计师到芝加哥、纽约、温哥华调研城市规划建设，特别是研究小地块的开发建设方式，仔细分析巴塞罗那、巴黎等城市街道的设计特点，这些经典案例给八大里规划指明了方向。下文以四里（四信里）、五里（五福里）为例，分析一下设计特点。

The Design Standards for Urban Residential Area Planning, which came into effect on December 1, 2018, adjusts the grading and scale of residential areas, refines the indicators related to land and buildings in residential areas, and optimizes the control indicators for supporting facilities and public green areas. People's behaviors are used as the standards to guide the planning of residential areas. In terms of traffic, the road network density should be enhanced and the road width should be reduced. The mix of residential and commercial areas and open communities are promoted. The route of people and vehicles is divided. The system rules to solve the problem of closed space. It developed the planning in the direction of people-oriented and provided the basic guarantee for scientific planning. The Badaili community in Tianjin, had the chance to practice the standard in Tianjin.

Tianjin Badali community planning was a government-led regional development project in 2013. At that time, the government leads the planning and the designers made add the details. At the beginning of the planning, to learn from foreign advanced urban planning concepts, they innovate planning methods. Decreasing the existing large residential areas and connect them. Make the living facilities meet the standard and be more accessible. Solve the problems of congestion, mess route of people and vehicles, and small and scattered green areas. Residential areas should be connected with other areas with different functions. Command organization let designers went to Chicago, New York, Vancouver to research urban planning and construction, especially for the study of the development and construction of small areas of land. They analyzed streets in Barcelona, Paris, and other cities. These case studies provide a pattern for the Badali planning. Here are some design features Sixinli community and Wufuli community.

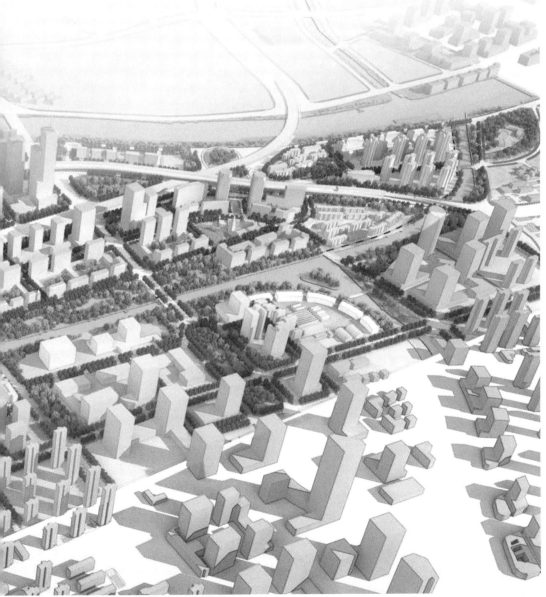

鸟瞰效果图 | aerial view of effect picture

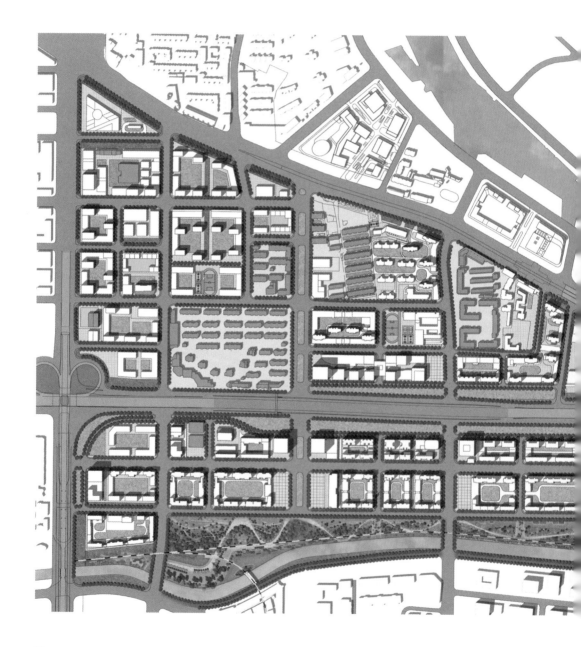

区位图 | location

总平面图 | general plan

窄路密网
Narrow Street · Dense Network

"窄路密网"的规划策略,需要考虑场地条件、市政条件及住宅规范的要求,结合房型经过反复推演,初步确定地块南北宽 80 m,东西长 150 m 为较合适的尺度。在这种尺度下,南侧可布置 8 层住宅,北侧可布置高层住宅,能满足日照要求,住宅面宽为 25 ~ 38 m,进深为 15 ~ 18 m,中央可以有 40 ~ 50 m 宽的院落。同时,四周由两层高裙房围合内院,四个方向设置出入口供应急消防车进入。按此标准地块布局,道路密度约 10 km/km²,达到了窄路密网的要求。

The strategy of enhancing road network density and reducing the road width is carried out. The designers consider the site municipal conditions and residential requirements, combined with the building style. It is a suitable scale for a site whose width is 80m from north to south and 150m from east to west. Under this scale, 8-story residential buildings can be on the south and high-rise residential buildings can be arranged on the north side. The width of residential buildings is 25m to 38m, the depth is 15m to 18m, and the central courtyard can be 40m to 50m wide. Besides, the inner courtyard is beside two-story-high podiums, and doorways are set in four directions for the fire exit. According to this standard, the road density is about 10 km/km², which meets the requirement.

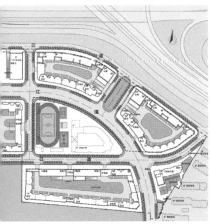

典型院落平面 | typical courtyard plan

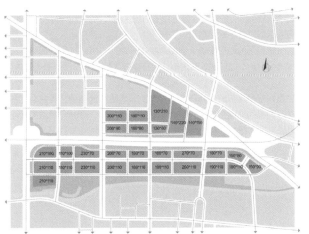

"七横七纵"的干道系统及院落平面尺度 | road system & courtyard plan scale

窄路密网效果图 | effect picture of narrow road and dense network

开放街区
Open Block

实现"开放街区"的规划,各地块的裙房主要是布置商业及配套设施,按照服务半径的要求规划配置各种公共服务设施,设施向街道开放,从街道可以直接进入从而可以避免穿入住区干扰居民生活。这样既保证了公共设施使用的便捷性,也保证了居民区的相对安静。创新地块边界设计,用地红线与道路红线统筹考虑,将人行道到用地红线及建筑根部区域按整体进行设计,增加建筑贴线率、提升步行区品质,增强街道的开放性与活力。

增加路网密度可疏解由网疏路宽带来的节点式交通拥堵,缓解交通压力,为组织交通提供更多的可能性。以"里"为单位进行内部的交通微循环,疏解城市主干道压力。同时,窄路密网小街廓的格局,促使市政道路紧邻用地穿越地块,为小地块带来便捷的公共交通等服务设施。

The planning is about an "open neighborhood". The podiums of each plot are mainly commercial and supporting facilities, and various public service facilities are planned according to the requirements. The facilities are open to the street and can be accessed directly from the streets, keeping more privacy for the residential areas. Such measure ensures both the convenience of public facilities and the relative quietness of the site. Also, they integrate the sidewalk to the red line of the site. The number of street stores is increased to enhance the openness and vitality of the street.

Increasing the density of the road network can relieve traffic congestion and provide more possibilities for traffic organization. Internal traffic micro-circulation by community can solve the traffic congestion on the city's main roads. At the same time, municipal roads are promoted to cross the plots, bringing convenient public transportation and other services to the small plots.

开放街区效果图 | effect picture of open block

混合住区
Mixed Community

实现"混合住区"。将地块内日照不足的区域布置为商业区及部分配套设施区,解决了配套服务及部分就业问题。同时,开发公寓类产品,保证居住产品供应的丰富性,满足不同群众对居住产品的需求。地块内商住比是4∶6,从区域角度对职住平衡起到积极促进作用,既缓解了开发商的资金压力,也为当地财政收入的永续打下了坚实基础。

Realizing the "mixed residential area" by arranging commercial functions and supporting facilities in the site with insufficient sunlight, which solves the problem of supporting services and partial employment. At the same time, we develop apartment products to ensure supply and meet the housing needs of different people. The ratio of commercial space to residential in the site is 4∶6. It not only leases the pressure but also makes the foundation for the local financial income.

沿公园一侧实景 | scene along the park

四信里住区商业办公面积与居住面积的平均比例为3：7，其中典型院落商业办公面积与居住面积的比例2：8

五福里住区商业办公面积与居住面积的平均比例为3：7，其中典型院落商业办公面积与居住面积的比例2：8

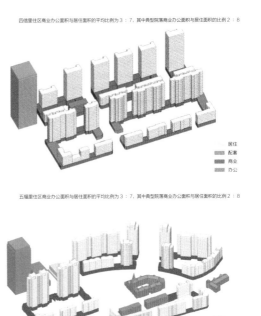

功能布局 | function layout

面积比例 | area ratio

绿地平衡
Green Space Balance

小地块带来绿化率不能满足"35%"硬性要求的弊病,规划采用集中绿地指标,给复兴河布局大型绿地,从区域角度实现整体平衡。所以,实现一个创新的规划从策划到实施,需要一整套的保障措施和具体的解决方法,尤其是在原有规划制度的体系下,更需要规划者具有超凡远见与卓越胆识。

Small green space is not enough. Our planning concentrates the green space to keep balance and meet the stander of the landscape ratio which is over 35%. Therefore, the realization of innovative planning requires a set of safeguards and specific solutions, especially under the system of the original one. It requires planners to have extraordinary foresight and courage.

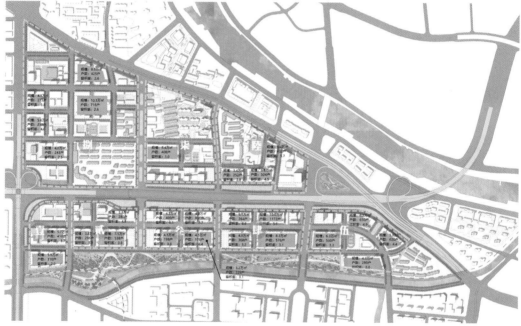

27个居住院落绿地系统 | green space system of 27 residential courtyards

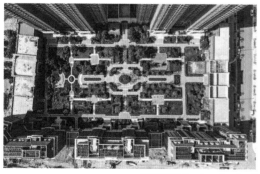

院落顶视图 | top view of courtyard

院落景观 | courtyard landscape

沿河景观 | landscape along the river

沿河一侧实景图 | scene along river

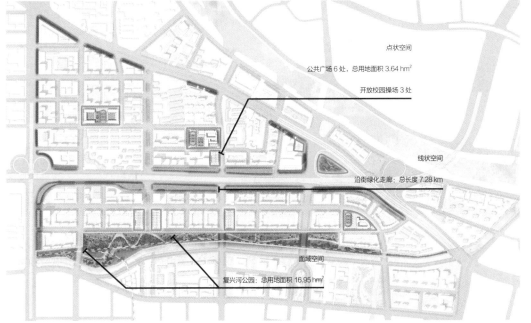

点状空间

公共广场 6 处，总用地面积 3.64 hm²

开放校园操场 3 处

线状空间

沿街绿化走廊：总长度 7.28 km

面域空间

复兴河公园：总用地面积 16.95 hm²

景观系统图 | landscape system

沿河景观 | landscape along the river

铁路景观小路 | railway landscape

院落环境
Planning Features and Courtyard Environment

基于以人为本和功能优先的设计理念，整个地块采用自然、简洁的设计手法，通过充分利用本地乡土植物，营造适宜生活、休闲和活动的社区环境，也为多样化风格的演绎提供了天然的平台。建筑围合成社区，在建筑院落空间延续宜居的品质，进一步打造便于邻里交流和适合老年人生活的院落环境。

Based on the design concept of people-oriented and function first, the whole plot adopts natural and concise design methods, and creates a community environment suitable for life, leisure and activities by making full use of local plants. The interpretation of diversified styles provides a natural platform. Under the conditions of surrounding communities created by buildings, we continue the pursuit of livable quality in the external space of buildings, and further create a courtyard environment that is convenient for neighborhood communication and adapt to the aging life.

宜居环境 | livable environment

院落顶视图 | top view of courtyard

院落鸟瞰图 | aerial view of courtyard

高度及界面控制
Height and Interface Control

多层级城市界面控制

Multi-level urban surface control

休闲界面：为沿公园打造活力休闲街，规划高 35 m 的连续街墙，配套设施和住宅，通过丰富的建筑立面展现浓郁的休闲风格。

Leisure surface: a vibrant leisure street along the park, with a street wall of 35meters in height, is a street for supporting facilities and residential.

活力界面：为创造都市商业街协调统一的风格规划，高度为 60 m 的连续天际线，混合多业态，建筑立面公建化，构建丰富的天际线。

Vibrant surface: the style of urban commercial streets is harmonized, with a skyline of 60meters in height. Building facades are publicized to build a rich skyline.

都市界面：沿黑牛城道保持 35 m 高的裙房线，沿路高层高度以 80 m 为主，重要节点高层高度控制在 100 ~ 220 m，打造城市底景。

Urban surface: 35 meters' building line along Heiniucheng Road, 80 m high rise along the road, and 100 m to 220 m high rise at important nodes to become the bottom view of the city.

高度控制 | height control

沿黑牛城道一侧实景 | scene along the Heiniucheng Road

道路一侧实景图 | scene along the road

沿铁道公园一侧实景 | scene along the railway park

值得欣慰的是八大里居住区先于《标准》而实施建设，并已陆续完成，尽管规划建设还有许多不足，但作为全国第一个"窄路密网、开放社区、混合住区"的实践者，按《标准》实施落地积累的经验与教训更值得我们去认真总结和深刻反思。

The great thing is that the practice of new standard residential areas in Tianjin has preceded the implementation of *the Standard* and has been completed recently. Although there are still many shortcomings in the planning and construction, as the first practitioner of "narrow roads and dense networks, open communities and mixed neighborhoods" in China, it has accumulated experience for the implementation of *the Standard*. It is worthy for us to summarize and to look back this practice.

整体鸟瞰图 | overall aerial view

2009年，电视剧《蜗居》带动"房奴"一词成为社会关注的焦点，"房奴"特指家庭月还款额占家庭总收入50%以上的家庭，因负债率高，已经影响了家庭正常的生活品质。

2010年住房和城乡建设部、国土资源部等七部委联合下发的《关于加快发展公共租赁住房的指导意见》是完善住房体系，培育住房租赁市场，满足城市中等偏低收入家庭基本住房需求的举措。该指导意见提出政府组织、社会参与、因地制宜、分别决策，统筹规划、分步实施的要求。天津市政府开展了保障性住房的规划建设工作。

项目选址在北辰区西部双青地区，距离北辰区政府4 km，距市中心小白楼地区15 km，占地3 km^2。规划有地铁1号线北延段和12号线。用地上，南北方向有北辰西道与津永路同周边联系，东西被南曹铁路与永清渠限制。场地边界三面有二级河道联通，内部水系丰富，植物覆盖率高。项目规划时充分考虑基地条件，结合天津"九河下梢""七十二沽"的自然风土，借鉴波士顿"翡翠项链"公园体系的概念，尊重自然肌理，提出营造"生态绿谷"的理念，同时提出"两河一谷，生态之城"的规划愿景。区域内西、北、东三面均有二级河道，利用内部既有水网结合地形高差以及地表植物分布，经梳理形成一条位于基地南部能够连通东西

双青新家园
Shuangqing New District

用地面积：302.55 hm²
建筑规模：2 900 000 m²

城市设计团队：
天津市城市规划设计研究总院建筑设计一院与规划设计一院

修建性详细规划团队：
天津市城市规划设计研究总院规划设计一院与建筑设计一院

获奖情况：
2013 年度天津市优秀城乡规划设计三等奖

鸟瞰图 | aerial view

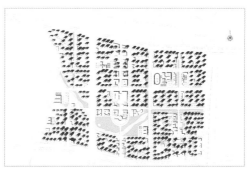

区位图 | location

河流的"生态绿谷"。绿谷占地约 25 hm²，长度为 2.2 km，宽度为 40～200 m 之间，它不仅给整个住区提供绿色休闲场所，同时串联各种功能形成复合型城市开放空间，促进城市中该地区水系的自然循环，也为在绿谷中生活的野生动、植物提供了良好的保育地。

空间布局上延伸生态绿谷形成三个片区，生态绿谷北侧以限价商品房为主，靠近永清渠的地块规划为商品房用地，生态绿谷南侧以公共租赁住房为主，公共租赁住房地块位于地铁等公共交通最为便利的区域，限价房和商品房也要满足交通出行、景观观赏等需求。空间上形成"一轴两区永续之城"的构架，中央绿轴布局配套公建项目。虽然分为三种类型居住用地，但公共服务配套设施、教育用地、绿地系统均按统一标准布局配建，形成区域商业服务中心和三级公共服务配套设施。

双青新家园在选址、规划、建设上均体现了"生态优先、以人为本"的先进规划理念，随陆续建成投入使用，得到各方高度认可。虽然规划是十年前完成的，当时的规划语境与今日已有所变化，但是其基本的规划理念、方法及实施效果都值得重新评估，若将其放在大的时代背景下重新审视，可以为将来城市发展建设提供丰富的经验。

天津北辰区　2011— 至今

In 2009, the term "house slave", which can be seen on TV shows, became the focus of our society. It refers to a family whose monthly debt payment is more than 50% of the total household income. Such situation has affected the quality of life of the family with a high debt rate.

In 2010, *Guidance on Accelerating the Development of Public Rental Housing* was launched as a way to improve the housing system, promoting the rental market and meet the needs of urban middle and low income families. It proposes government organization, social participation, should separate decision-making and integrate planning step-by-step. The Tianjin Municipal Government carried out the planning and built guaranteed housing.

The project is in the western Shuangqing area of Beichen District, 4 km away from Beichen Government and 15 km from the Xiaobailou area in the city center. It covers 3 square kilometers. There are two subway lines No.1 and No.11 linked to the site. The site is between Beichen road and Jinyong road from the south to the north and connected to Nancao railway and Yongqing Canal. There are three river channels on different sides, which provide rich water resources and increase plant coverage. In the planning of the project, we considered the basic conditions combined with the culture of rivers in Tianjin. With the concept of "Boston Necklace", the planning of protecting natural texture, building eco-valley, and making the two rivers and the valley the environmentally friendly city was proposed. There are secondary rivers in the west, north, and east of the site. The varied topography and water network provide an enabling environment for the plants. It is made to form an "eco-valley" in the south that can connect the rivers in

方案总平面图 | general plan of the scheme

与城市整体区位关系 | relationship with the overall location of the city

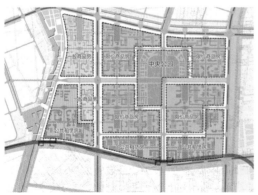

功能分区 | functional partition

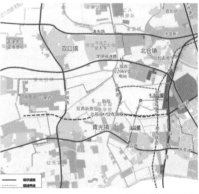

四至范围 | plot range

the east and west. The eco-valley covers about 25 hectares, with a length of 2.2km and a width of 40 to 200m. It not only provides entertainment areas for the local people, but also links various functions to build a compound urban public space. Besides, it promotes the natural circulation of the water system in the city and provides a good conservation area for wild animals and plants.

The Eco-Green Valley was extended to form three zones: the low- price housing zone in the north, the commercial housing zone near the Yongqing Canal, and the public rental housing area in the south. The public rental housing area is the most convenient area for public transportation. The low-price housing and commercial housing ensure the whole neighbour meets the needs of transportation and landscape. Spatially, the site forms a structure of "one axis and two zones of a sustainable city". That means the central green axis of the city is surrounded by supporting public buildings. Although there are three types of residential areas. The public service facilities, education land, green space system, and all are laid out and built according to a certain standard, forming a regional commercial service center.

The site selection, planning, and construction of Shuangqing New District show the advanced planning concept of ecological priority and people-oriented, which have been highly praised. Although the planning was completed ten years ago when the enviroment was not the same as today. The planning concepts, methods, and implementation are worth valuating and examining. Such planning can collect experience for future urban development and construction.

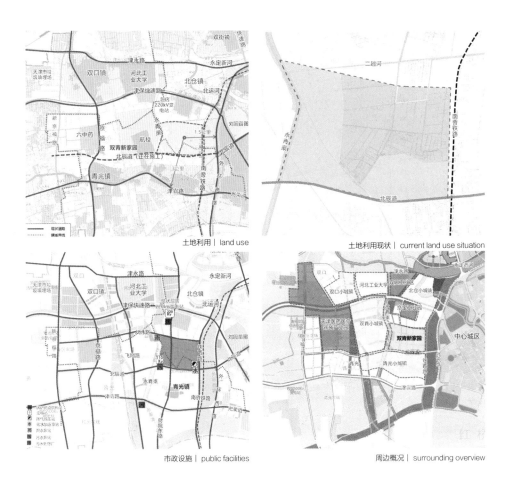

土地利用 | land use

土地利用现状 | current land use situation

市政设施 | public facilities

周边概况 | surrounding overview

规划设计比选方案
Scheme Comparison

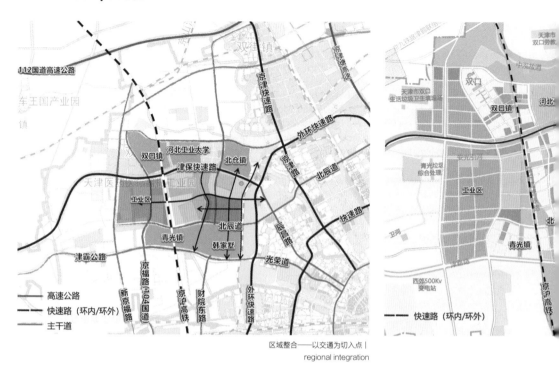

区域整合——以交通为切入点 |
regional integration

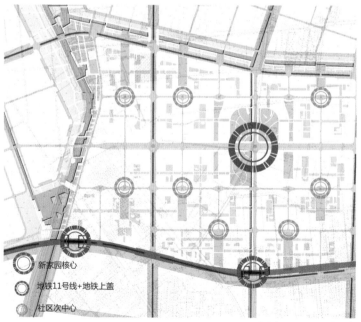

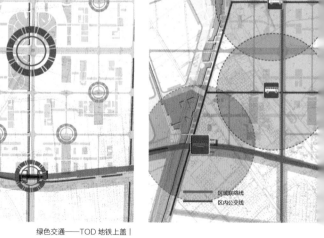

绿色交通——TOD 地铁上盖 |
green traffic—TOD subway building over the track

区域整合——构建三横三纵路网体系 |
regional integration

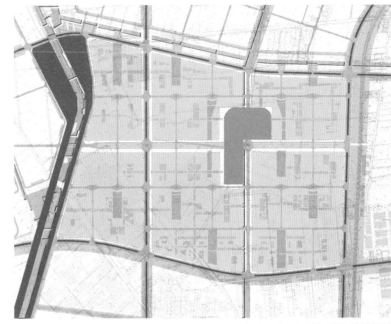

生态保护——保留景观核心，连接两侧生态绿地 |
ecological protection

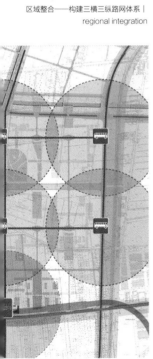

绿色交通——区域内公交系统 |
traffic—public traffic system inside the area

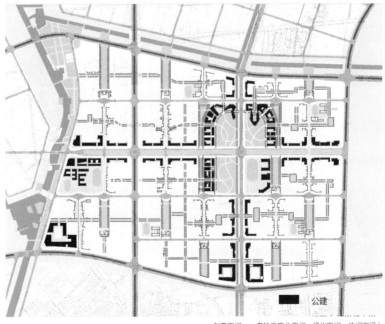

创意空间——多种类商业空间、绿化空间、休闲空间 |
creative space—multiple commercial space, green space, leisure space

规划设计选定方案
Selected Scheme

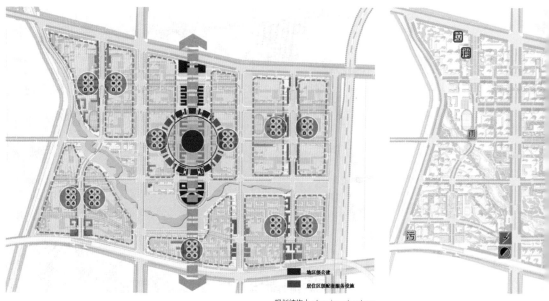

规划结构 | planning structure

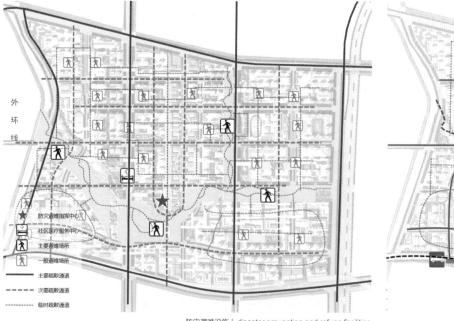

防灾避难设施 | disaster prevention and refuge facilities

一轴：综合配套公建服务轴。一带："生态绿谷"景观带。六片区：六个居住片区。十一邻里：十一个居住邻里。

One axis: comprehensive public building service axis; one belt: "Eco Valley" landscape belt; six zones: six residential zones; eleven neighborhoods: eleven residential neighborhoods.

市政设施 | public facilities

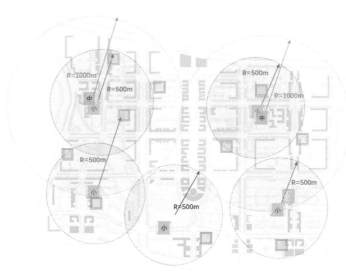

公共服务设施 | public service facilities

慢行系统 | slow traffic system

绿化系统 | greening system

开放空间
Open Space

中心开放空间——一级带状生态湿地体系 |
central open space-class Ⅰ—ribbon wetland ecosystem

邻里开放空间——二级组团绿地 |
neighborhood open space class Ⅱ—group green space

联络空间——三级慢行系统相连 |
contact space class Ⅲ—connected slow-vehicle system

开放空间系统 | open space system

公共服务设施
Public Facility

一级配套公建——连接南北两侧居住区，避开主要干道 |
class I public buildings—connect the communities along the street and keep away from the trunk roads（包括学校、医院、养老院、银行等区域级公共服务设施）

二级公建——组团服务 | class II public buildings—group services
（包括幼儿园、便利店、健身房、棋牌室、储蓄所、邮局等公共服务设施）

三级公建项目——沿街基础服务设施 | class III public buildings—infrastructure along the street
（包括沿街便民店、洗衣房、早点铺等最基本的生活服务项目）

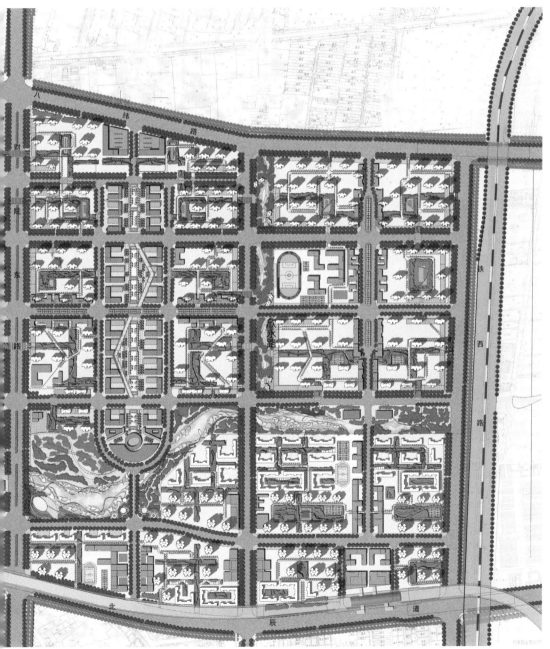

总平面图 | general plan

带状湿地公园
Strip Wetland Park

带状湿地公园的中心与中央配套公建轴的建筑意向相呼应,引导居民进入公园,同时在公园中设置景观停留点,例如:浮桥、观景平台等活动场所。

The center of the strip wetland park is similar in architectural style to the central public building axis, guiding residents into the park, while setting up landscape stopping points in the park, such as pontoons, platforms, and other activity places.

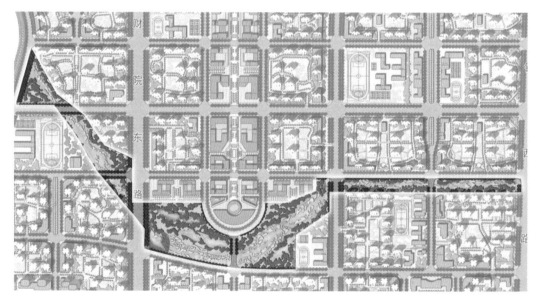

带状湿地平面图 | strip wetland park plan

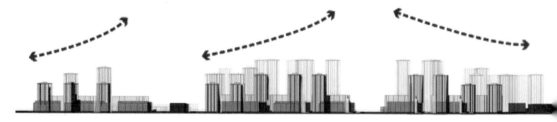

沿组团配套公建项目两侧的居住区天际线不会压迫道路　　　　　　　　沿中央公园两侧

带状湿地效果图 | effect picture of strip wetland park

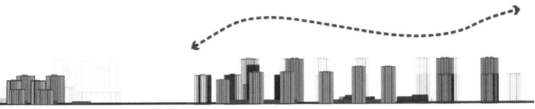

由低渐高的态势　　　　　沿公园一侧的限价房区域天际线变化丰富　　　　　天际线 | skyline

中央配套公建轴
Central Public Building Axis

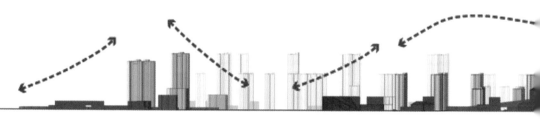

沿中央公园两侧天际线高度呈现出由低渐高的态势

中央配套公建轴在整个地块的核心位置，为市民的日常活动和商业活动提供了一个良好的场所，同时中央配套公建轴和湿地公园的良好结合，使居住区的生活节奏在此发生改变。

The central public building axis is at the core of the whole site, providing a good place for the daily needs of citizens and commercial activities. While the combination of the central service axis and the wetland park makes the lifestyle change here.

中央配套公建轴效果图 | effect picture of central public building axis

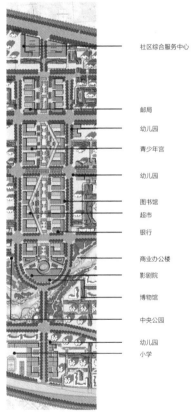

社区综合服务中心
邮局
幼儿园
青少年宫
幼儿园
图书馆
超市
银行
商业办公楼
影剧院
博物馆
中央公园
幼儿园
小学

中央配套公建轴布局 | layout of central public building

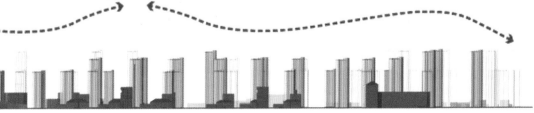

中央配套公建轴两侧的天际线变化丰富

天际线 | skyline

双青新家园建成图
Photo of Completed Shuangqing New District

建成实景图 | scene

义乌是个神奇的地方，乘改革开放东风，义乌人将义乌这个"鸡毛换糖"的地方建成东方小商品圣地，当地人与时俱进、勇于自我革新的精神，引领他们向更高目标迈进。承接义乌项目，同义乌结缘，对于很少在外地开展工作的我们来说这是个挑战，好在几年的奔波都坚持下来了，回头看看义乌异国风情街的复兴还是特别欣慰的。

来到义乌之后，设计团队先匆匆看了一下现场，除了建筑密度大、环境杂乱之外，没有留下什么特别的印象，但在参观完义乌小商品市场之后，我们对琳琅满目种类齐全、设计精美的小商品叹为观止，同时，传统印象

义乌异国风情街
Yiwu Exotic Street

用地面积：26 hm²
建筑规模：330 000 m²

城市设计团队：
天津市城市规划设计研究总院

修建性详细规划团队：
天津市城市规划设计研究总院

合作团队：
天津华汇工程建筑设计有限公司
天津市博风建筑工程设计有限公司
天津中天建都市建筑设计有限公司
天津大地天方建筑设计有限公司

获奖情况：
2019 年度全国优秀城乡规划设计奖（城市设计类）三等奖
2019 年度天津市优秀城乡规划设计奖（城市设计类）一等奖

浙江义乌市　2016—2018

鸟瞰图 | aerial view

区位图 | location

中小商人粗制滥造、精于算计的印象也被颠覆。我们为"世界小商品之都"的实力所震撼，这一切不仅改变了我们对小商品的认识，也改变了我们对义乌人的印象，更让我们从内心重视该任务，并倾注全部精力与热情。

城市更新的案例不少，但成功的例子不多。我们团队谨慎地开始了前期工作。在前期调研工作中发现，义乌的小商品从业者有很高的参与意愿，同时当地政府也比较开明，许多改造都是来自上至下的引导和推动。

113

Yiwu is a magical place. Since the Economic Reform and Openning up, Yiwu people have changed Yiwu from a small marketplace to the mecca of small commodities in the East. Their courage of innovation makes them achieve higher goals. It is a challenge for us who seldom work in another city to undertake the project. However, we hang in there. Now, it is gratifying to look back at the revival of the Yiwu exotic street.

In Yiwu, our first thought of the site was unimpressive, except for the dense buildings and the cluttered environment. However, after visiting the small commodity market, we were shocked by various small commodities with exquisite designs. At the same time, the stereotype of small businessmen who are calculating and greedy was overturned. We were shocked by the consumption of products. It not only changed our understanding of small commodities but also changed our impression of the Yiwu people. It made us pay attention to the task and fully commit to the project.

There are many cases of urban renewal, but few of them are successful. In most cases, the vibrant economies would lose their features after the renewal. Since that, our team did not take any concept of implantation but started researching. In the research, we found that Yiwu's businessmen have a high willingness to participate. The government is also more enlightened. Many of the previous planning decisions were made by the local people.

小商品市场发展历程
Development History of Commodity Market

挑货郎 | peddler   湖清门小百货市场 | Huqingmen Commodity Market   新马路市场 | New Road Market

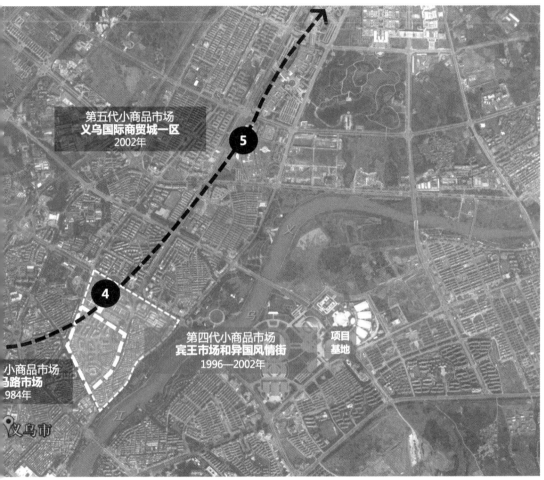

小商品市场发展历程 | the development of commodities market

城中路市场 | Chengzhong Road Market 　　　　义乌宾王市场 | Yiwu Binwang Market 　　　　义乌国际商贸城 | Yiwu International Trade Town

现状认知
Development History of Commodity Market

经过二十几年发展，作为第四代小商品市场的宾王市场开始衰落，街道设施破旧，设备年久失修，业态杂乱无章，传统的小商品商业已经失去活力。需要新业态导入、商业布局优化、空间情境再造等更新工作。

After twenty years of development, the Binwang market, as the fourth generation of commercial space began to depress. The streets are dilapidated, the equipment is in disrepair, and the traditional products have lost their appeal. New businesses need to be introduced into the commercial layout. In that way, it can be a vibrant space to implement the renew.

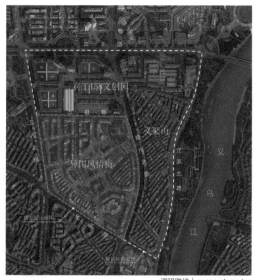

调研路线 | research routes

调研照片 | research photos

现状认知
混合社区
居住 + 商业 + 服务

现状照片 | current situation

社区管理 | community management

功能业态 | functional state

空间环境 | space environment

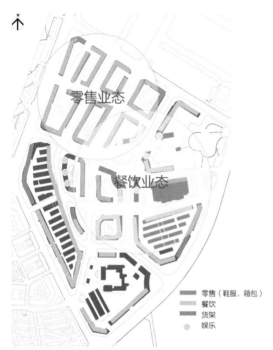

在业态平面布局方面，餐饮业态在空间上主要在街区南部集中，零售空间则集中在街区北部。

In terms of the layout of the business plan, the restaurant business is concentrated in the south of the block. The retail is on the north of the block.

业态分布图 | the business pattern map

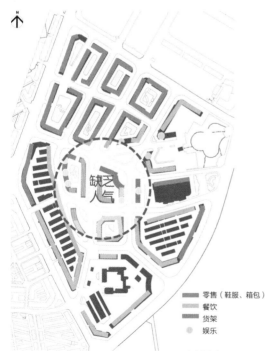

未来设计团队结合道路体系及社区运行机制，导入人流、优化业态，打造街区中部重要节点。

In the future, the road and community system will bring in people and optimize the business, forming a significant node in the middle of the neighborhood.

业态分析 | business analysis

现状分析 | present situation analysis

社区 | community

经济 | economy

环境 | environment

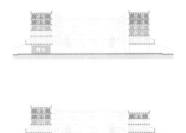

产居商三合一 高度混合
空间形态"四层半"

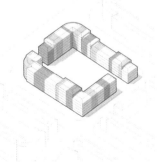

院落产权 垂直划分
垂直户数 29 户

建筑平均层数 5.4 层
建筑密度 42%

建筑形态 | architectural form

注：过剩，指当地人力物力过剩。过客，指当地有很多外来人员。过渡，指当地有很多空间缺乏管理。

结合社会管理、经济结构、街区环境提出"三变三不变"。"三变"是产业结构升级、经商环境提升、经济效益改善,"三不变"是营商模式不变、产权归属不变、建筑布局不变。在与居民的交流沟通中,政府部门将改造中的"变与不变"传达给居民,并建立起"自上而下地指导,自下而上地参与实施"的机制,形成"多方位、创意强、低冲击、微改造、渐进有机更新"的义乌模式。

Considering social management, economic structure, and neighborhood environment, we proposed the "Three changes" and three unchanged." "Three changes" are industrial structure upgrading, commercial environment upgrading, and economy increasing. The "three unchanged" are keeping the business model, property rights, and street-store relationship unchanged. The core of the renew and stability was passed on through communication with residents. The policy was made by the planner and guided by the local people. The new Yiwu model was built in various dimensions and with low-transformation and high innovation.

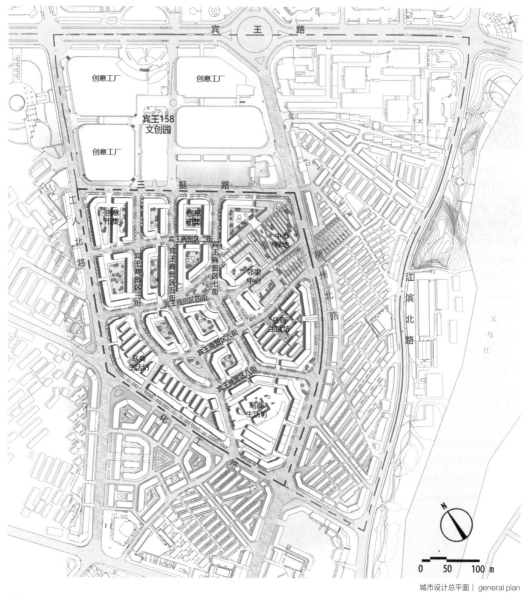

城市设计总平面 | general plan

三变三不变：三变 | Three Changes

| 产业结构升级 | 街区环境提升 | 经济效益改善 |

改造前 | before transformation

收益上涨幅度：15~36%

改造后 | after transformation

三变三不变：三不变 | Three Unchanged

| 营商模式不变 | 产权归属不变 | 建筑布局不变 |

改造前 | before transformation

改造后 | after transformation

【机制创新】

社区共管共治
社 区 自 治 的 制 度 建 立

成立"社区议事会"

1 在社区商业管理委员会、商户协会框架下,建立社区议事会,订立议事会制度,成员由业主与租户(商家)代表共同组成。

2 议事会制定并实施社区公共服务基金的收取办法。根据社区实际情况,依业态与建筑面积收取相应的社区公共服务费。

3 议事会成员商讨并决定基金的具体支出。结合社区实际情况,商讨使用范围及支出具体数额,合理使用基金。基金使用采取"一事一议"原则,也可分年度制定支出计划。

4 社区议事会向业主与租户(商家)报告基金的收支情况,接受全体业主与租户的监督。

【市场】

所有收支费用需公示并接受政府、社区、协会及社会的监督

集中社区管理费,形成社区持续、有序更新的资金支持

形成更新示范效应

基于区域重商传统及商会组织
优化商业运营机制
营造更优越的商业氛围

【政府】

政府主导下的城乡新社区投资集团和城市建设投资公司

遵照《义乌市关于进一步加强和规范物业管理工作的意见》(2014年)合法、合理、公开透明收取物业管理费

提高城市综合治理能力

商业管理委员会
由所有业主构成
管理社区内的生产资料

【社区】

建立社区商业管理的主体
对于社区、商业运行进行调控

通过社区激励、奖励评比等方式
鼓励居民"自下而上"地参与社区更新

形成"自上而下"与"自下而上"
相结合的社区更新机制

商户协会
由所有经营户组成
维护经营秩序与权益

共谋 — 共筹 — 共建 — 共管

首先是尊重居民参与意愿,强化社区归属感。按照社区共管共治的制度建立社区议事会,依托政府公信力,收取社区公共服务费用,维护正常运作,同时筹集社区发展资金,服务社区建设。商户通过社区议事会与设计人员建立沟通,以开放包容的态度,重塑社会交往的公共空间载体,双方对社区更新形成共识与合力。

The first is to respect the residents' willingness to participate and strengthen the sense of belonging to the community. The community council is established according to the system of community co-management and governance. Relying on the credibility of the government, the community council can raise funds, maintain operation, and serve community construction. Merchants communicate with designers through the community council to reshape the public space for social communication with an open and inclusive attitude. It proceeds the renew of the society.

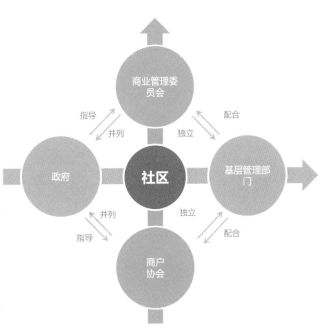

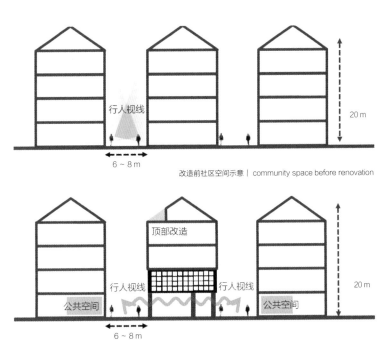

改造前社区空间示意 | community space before renovation

改造后社区空间示意 | community space renovation

外源性人群 导入		内生性人群 成长

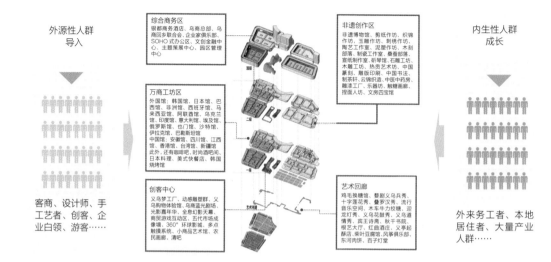

客商、设计师、手工艺者、创客、企业白领、游客……		外来务工者、本地居住者、大量产业人群……

其次，尊重市场发展规律，推动产业持续更新。引入文创资源助力互联网创新创业，提倡业态的混合多元。在维持竖向分层排布功能的基础上，底层业态进化为由市场自然引导。顶层引入创意设计、线下体验等孵化产业，适时选择落地（向下层迁移），为区域小商品的交易提供适应市场需求的消费方式，同时增加配套服务，提升综合服务水平。

Second, it is important to respect the law of market development and promote industrial renewal. Introduce cultural and creative industries to help Internet industries innovate, and advocate the mixed various of industry. On the basis of maintaining the vertical arrangement, the natural evolution of the bottom business is guided by the market. The top layer has creative design space, offline experience and other industries. They also migrate to the lower layer to provide suitable space for the trading of the small commodity trade. Besides, we also increase supporting services to enhance the level of comprehensive services.

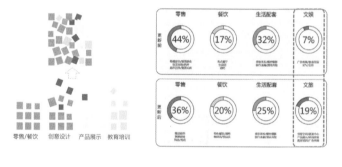

垂直业态 | vertical business

124

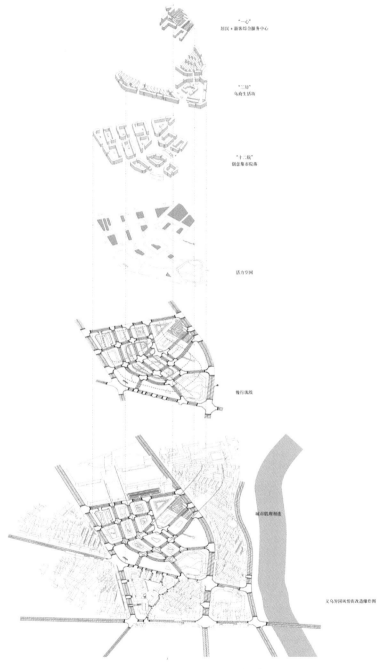

爆炸图 | exploded diagram

最后，尊重城市印记，重塑表里如一的空间环境。主街结合骑楼的设置，使购物空间尺度更宜人，生活街道的连续让购物体验更舒适，改变内院人车混行、公共空间被侵占而造成的利用率低的现状，实现了人车分流，将普通绿化地提升为开放共享的积极空间，重新为其赋予活力。

Finally, we respect the city's feature and reshape the environment as it required. The main street is combined with the additional arcade building to make the scale of the shopping space more pleasant, the continuity of the living street makes the shopping experience more comfortable. It changes the mixing of people and vehicles in the inner courtyard and the low utilization rate of the public space, separates the route of people and vehicles, and upgrades the common green space into an open space with vitality.

改造工程与现有建筑有机结合，保留街区特色

整合屋顶及檐线，优化店铺立面

以模块化的方式整合现有立面

统一店铺招牌位置、规格，活跃店铺招牌样式

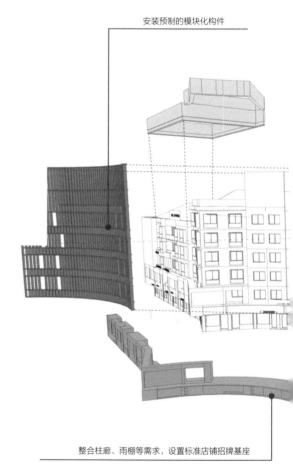

安装预制的模块化构件

整合柱廊、雨棚等需求，设置标准店铺招牌基座

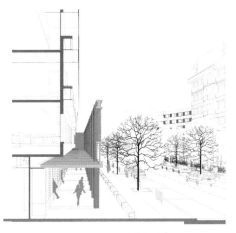

外廊剖面图 | section of gallery

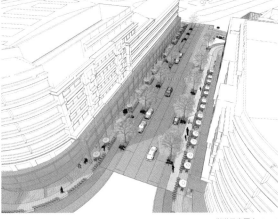

街道示意图 | street

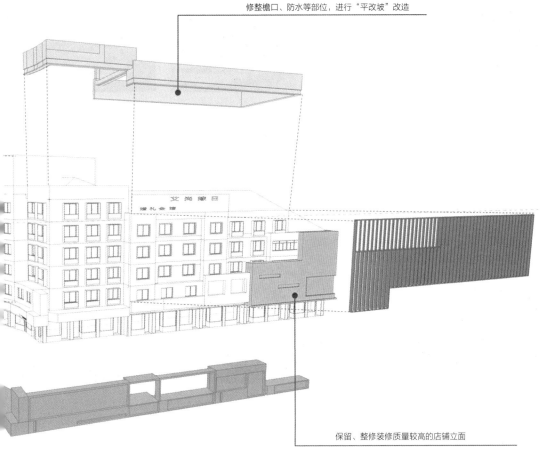

修整檐口、防水等部位，进行"平改坡"改造

保留、整修装修质量较高的店铺立面

具体改造做法 | specific transformation practices

改造前后对比 | Before & After

改造前：宾王商贸区三街 |
before transformation: No.3 street of Binwang Business Trade District

改造后：宾王商贸区三街 |
after transformation: No.3 street of Binwang Business Trade District

改造前：F-01-01 号楼 |
before transformation: No.F-01-01 Building

改造前：B01-01 号楼 |
before transformation: No.B01-01 Building

改造后：F-01-01 号楼 |
after transformation: No.F-01-01 Building

改造后：B01-01 号楼 |
after transformation: No.B01-01 Building

经过新形势下的规划策划,城市更新改造逐渐实施完成。随着改造逐渐完成,老旧街区正逐渐成为一个引领义乌新风尚的文旅体验区,传统商业街衍生出新的业态。同时,它渐渐散发出向外辐射的能量,三挺路活跃的空间经济使更新后的异国风情街成为义乌江畔的活力街区,在新经济浪潮中成为"全民参与、低成本、微冲击、渐进式"城市更新模式的成功典范。

After city renewal in the new situation, the old neighborhood becomes a culture-oriented tourism experience area, which leads to a new fashion in Yiwu. The traditional commercial street has given birth to new business models. At the same time, it has the potential to affect other places. The active space of Santing Road makes the renewed exotic street a full-time dynamic neighborhood along Yiwu River. It has become a successful model of universal participation with the low-cost and gradual progress of the urban renewal model.

鸟瞰图 / aerial view

天津这座城市因河而兴,古称天子渡口,特别是清末九国租界的设立,对海河两岸的城市格局影响至今,从北部意租界、日租界到中部德租界、俄租界再到西南英租界、美租界,它们被动地勾勒出天津近代的城市版图。民国期间随近代工业落地海河边,热电厂、棉纺厂等近代工业设施逐渐建成。2004 年,海河改造带动了天津新一轮城市发展,租界得到了很好的保护,但老厂区却没有那么幸运。我们接手第一热电厂地块改造的时候,只剩下当初建设的中央厂房,但其有代表性的烟囱已经被粗暴地拆除,踪影难觅。

海河沿线的开发已过去十几年,当初的建设在改变城市的同时也留下了些许遗憾,一些历史建筑一旦被拆除或损毁将很难复原。这些教训提醒后来人要尊重老建筑,开发建设时应对其进行保护性更新,在保护中开发,在更新中发展。

天津市第一热电厂
Tianjin No.1 Thermal Power Plant

用地面积：13.75 hm²
建筑规模：373 000 m²

城市设计团队：
天津市城市规划设计研究总院

鸟瞰图 | aerial view

区位图 | location

接手项目后，我们秉承有机更新的原则，提出保留老厂房及前面至河岸的绿地，集中开发周边区域的总体规划方案。我们考察天津市内海河沿岸后发现，除了津湾广场（原英租界、法租界）留有一个滨水广场作为城市开放空间接纳从天津站来津的游人之外，海河沿线只有几个小型的游船码头，分别位于古文化街、利顺德大饭店等处，除此之外再无开放的岸线，这处绿地给已经过于压抑的海河沿岸留下了仅有的喘息的空间。第一热电厂老厂房前绿地成为海河沿线打造滨水公共空间的唯一区域，因此我们的方案重点是保住老厂房及周边环境，留下场地的记忆，保留绿地丰富沿岸景观，将丰富的河岸空间还给广大市民。

天津河东区

2011至今

Tianjin is built by the river, which was called the Emperor Ferry in ancient times. The construction of concessions in the late Qing Dynasty influenced the urban planning on both sides of the Haihe River till today. From the northern Italian and Japanese concessions to the central German and Russian concessions to the southwest British and American's, they outlined the modern urban planning of Tianjin. During the Republic of China, industrial factories, such as the thermal power plant and the cotton spinning factory, were built along the Haihe River. In 2004, the Haihe River renovation increased the development in Tianjin, and the concessions were well protected. However, the old factory areas were not so lucky. When we took over the site, only the central plant left by the Japanese was still there. The chimney which is the most representative had been roughly destroyed.

Ten years have passed since the development along the Haihe River, and the demolition and new construction beautified the city, but also left the new city with regrets. It reminds people to respect the old buildings. Construction should give way to protection and restoration. Urban planning should be with conservation and renewal.

After taking over the project, we proposed a master plan to preserve the old factory buildings and the green space in front of them to the river bank. We concentrate on the development of the surrounding area according to the principle of city regeneration. Except for a waterfront plaza at Jinwan Square (former British-French Concession), an open space for visitors from Tianjin Station, there is no open space other than a few small ferries along the river located at Ancient Culture Street and Astor Hotel. This green space is the only open space along the dense Haihe River. The green space in front of the old thermal power plant buildings is the only chance to create a water-accessible public space along the Haihe River. Therefore, we need to focus on preserving the old factory buildings and the green space and enriching the landscape along the river. In that way, we can recall the memories of the site and provide a place for people to enjoy lives.

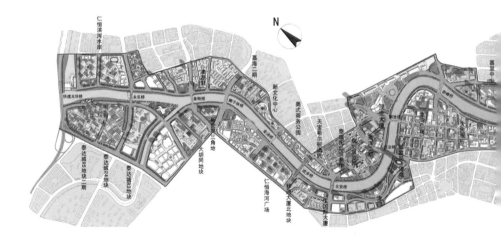

厂房改造前 | before transformation

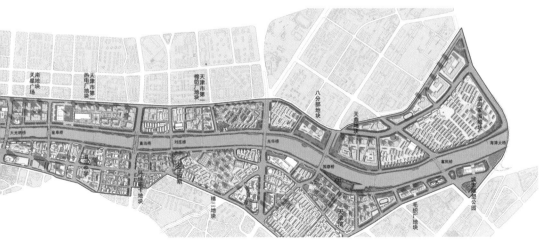

海河长卷 | the painting scroll of Haihe River

确定场地的核心价值之后，其他地块就围绕着老厂房来统筹布局。海河沿岸的地块采用围合院落的方式设计洋房产品，将高度限制在与老厂房保持一致的范围之内。后面的地块，采用高层住宅与裙房结合的方式处理，沿六纬路一侧布置商业区与城市连接，中央地块在地铁出入口周围建两栋高层写字楼，一是充分发挥以公共交通为导向的开发模式（TOD）的效力，二来可以平衡拆迁成本，三是为河东区带来永续的税收收入。

After determining the core of the site renovation, the other plots were designed around the old factory buildings. Along the Haihe River, houses are built in the style of enclosed courtyards, and the height is limited to the same as the old factory buildings. There is a combination of high-rise residential buildings and the annex to the main building along Sixth Weft Road, which increases commercial development in the city. Two high-rise office buildings in the center with the subway entrances and exits balance the cost and bring tax income for Hedong District.

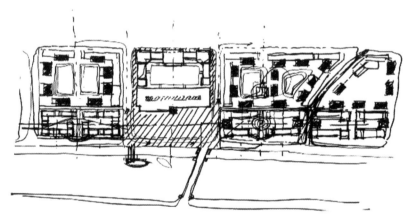

总平面草图 | general plan sketch

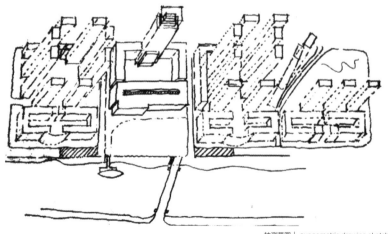

轴测草图 | axonometric drawing sketch

规划布局以"一个公共化园、两个居住组团、二个商业院落"为原则,建筑群沿海河成组布置,空间关系逐层升高。

模型图 | model

地块规划统一考虑周边地块布局，形成整体空间效果。规划布局以"一个公共花园、两个居住组团、三个商业院落"的原则进行，保证建筑群沿海河成规模，成组布置，且空间关系逐层升高。

The planning of the site takes the layout of the two surrounding areas into account, which sets the overall effect. The planning layout is under the principle of "one public garden, two residential groups, and three commercial compounds", ensuring that the buildings are arranged in groups along the river and that the spatial relationship rises level by level.

总平面图 | general plan

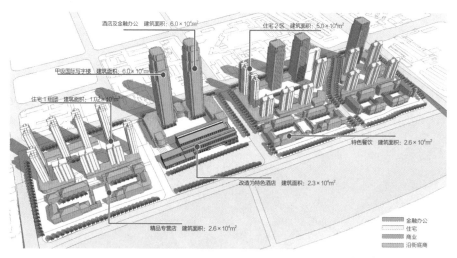

业态比例 | business proportion

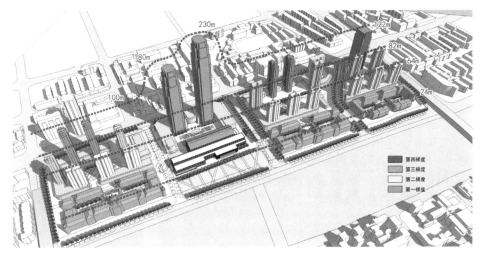

高度分析 | height analysis

鸟瞰图 | aerial view

临河院落 | riverside courtyard

内院 | inner courtyard

商业内街 | commercial inner street

海河沿岸效果图 | effect picture along Haihe River bank

内部场景 internal scene

现场勘查老厂房时给我们留下深刻印象的是厂房内的梁柱都保存完好。虽历经百年沧桑,但坚固依旧。站在老厂房旁边就好像停留在一个被过去的时光包裹的大空间里,整个场地都给人梦幻之感,使人感动不已。这种感受坚定了我们将它保留下来,同时将它与海河间的场地作为城市开放空间的决心,这样做既可以从城市层面解决海河沿线千篇一律、单调乏味的问题,也给海河沿线提供了弥足珍贵的开放空间,还可以达到保护老厂房及其周围环境的目的,从单纯物理意义上保护老厂房,提升到保护它存在的环境而对其进行系统地保护和利用,实现真正的有机更新。

We were impressed by the well-preserved beams and pillars of the old factory. Despite a hundred-years, suffering earthquakes and floods, they are still sturdy and can still support for many years. We could not help but marvel at the construction skills and quality of the time. Standing next to the old factory building, it is like staying in historical space The whole site gives you a sense of dreaming back to the past and touches you. This feeling made us decide to preserve it. The room between it and the Haihe River is used as a public open space. This can provide precious open space along the Haihe River, and preserve the old factory building and its surrounding environment.

2012 年厂房照片 | plant picture in 2012

厂房外景效果 | external effect picture

沿河岸立面图 | business

商业街景图 | business street

后来该地块被金茂集团买下，其开发建设严格按照我们的城市设计开展，预计 2022 年完成。老厂房被改造为社区活动中心，集酒店、餐饮、娱乐于一体。绿地被设计为一个向市民开放的公园，相信将来在海河边会出现一个不一样的空间，希望它能给后人带来像法国塞纳河左岸一样的生活体验，它承载着我们保护海河、复兴天津的愿望、祈盼和理想。

Then, Jinmao Group won this project and carried out the construction according to our urban planning. It will be completed by 2022. The old factory buildings are transformed into a community activity center with a hotel, restaurants, and other projects. A park is designed for the public. We believe that a different place will appear in the future by the Haihe River and we hope it will provide a bank lifestyle for future generations like the Seine River in France, which carries our desire and dream to protect the Haihe River and revive Tianjin.

海河沿岸效果 | effect along Haihe river

西于庄位于天津老城厢以北，紧临子牙河，在北运河和海河的交汇处。随天津西站的建成并投入使用，位于西站北侧腹地的西于庄的价值日益显现。西于庄在历史上是依大清河南岸自然形成的小村落，宋辽时属武清管辖，后几易其名，近代方定名为西于庄并沿用至今。其倚中环、邻西沽、接北辰、与西站隔河相望，地理位置重要。金代漕运把三岔河口定为直沽寨，子牙河是三岔河口的重要水道。清代这里水运发达，漕运兴盛，一年从开河到封河有近两万艘船路过，资料显示当时的西于庄是清代运河上的重要物流中心。

西于庄老城区始建于20世纪50年代，由于年代久远而民居危陋，年久失修，街巷杂乱无章，出行不便，是天津最大的棚户改造区（以下简称棚改区），被市政府列入整体拆迁区域。该棚改区位于三河交汇处，占地64 hm²，北侧有老北洋大学（现河北工业大学）。2013年启动拆迁并开展规划工作，当时正值《天津市城市总体规划》修编，经专家论证，在统筹考虑天津社会、经济、政治等因素之后，将西于庄地区定位为天

西站西于庄
Xiyuzhuang Area of Tianjin West Railway Station Business District

用地面积：64 hm²
建筑规模：1 250 000 m²

城市设计团队：
天津市城市规划设计研究总院

合作团队：
天津大学建筑设计规划研究总院
张开宇工作室

获奖情况：
2021 年度优秀城市规划设计奖

鸟瞰图 | aerial view

区位图 | location

津城市副中心之一，并按照 CBD 进行规划定位。天津主中心为小白楼地区，西于庄地区和柳林天钢分别为两个副中心。

对于依托西站老城区的发展，不仅要考虑其所在区域红桥区的条件，更要综合分析天津以至周边区域的情况。2014 年中央提出将"京津冀协同发展"列为国家战略，西站作为天津的重要交通枢纽，连接北京、天津、河北，特别是雄安新区，在区域联动方面发挥了巨大作用。目前京津冀已初步实现高铁层面上的互联互通。

2019 年，随西于庄拆迁工作的彻底完成，市政府随即按照西于庄规划方案对该区域规划进行进一步设计。新时代新需求，天津城市建设经历十几年的高速发展后，现在进入注重品质、保护传统、关注民生、保护生态的新阶段。

Xiyuzhuang is in the north of Tianjin Old Town, close to Ziya River, beside the North Canal and Haihe River. With the operation of Tianjin West Railway Station, the value of Xiyuzhuang, which is the hinterland of the Tianjin West Railway Station, has increased. Historically, the Ziya River was called the Grand Canal, under the control of Wuqing during the Song and Liao Dynasties. This area was later called Xiyuzhuang. It is next to Zhonghuan, Xigu, Beichen District and across the river is the Tianjin West Railway Station. In the Jin Dynasty, the government made estuary area the base of Zhigu. The Jingbao Company in 1900 was later the Hebei Provincial Shipping Bureau. At that time, water transportation thrived so the business on the canal was quite important. Nearly 20,000 ships passed by the canal in a year. Xiyuzhuang was so significant in the Qing Dynasty.

The old town of Xiyuzhuang built in the 1950s is now dangerous and disorganized. It is the largest rickety cottage area in Tianjin and needs to be demolished and relocated as a whole. The current town is located at the crossing of three rivers, covering 64 hectares, with the old Beiyang University (now Hebei University of Technology) on the north side. In 2013, *The General Plan of Tianjin* planned the Xiyuzhuang area as one of the sub-centers of Tianjin after taking into account the social, economic, and political factors of

Tianjin. Tianjin would have the main center Xiaobailou area, and Xiyuzhuang area and Liulin Tiangang as two sub-centers. Thus, the construction of the western Tianjin CBD became the development target. The 2015 proposal put the scale of development in the first place, providing the government with a solution to meet the demands of enterprises and residents.

For the development of Tianjin West Railway Station, the planner should not only consider the conditions of the Hongqiao District, but also measure the overall condition of Tianjin. The central government put forward the Beijing-Tianjin-Hebei integrated development plan in 2014. The West Railway Station is a significant hub of Tianjin. The high-speed railway station connecting Beijing, Tianjin, Hebei, especially the Xiong'an New Area, is playing an important role.

As time went by, the demolition and relocation of Xiyuzhuang finished in 2019. The revision and upgrading of Xiyuzhuang's planning scheme would be carried out. After more than a decade of high-speed development, Tianjin's urban construction focus more on quality, preserving traditions, paying attention to people's livelihood and protecting ecology to meet the need for the new era.

透视图 | perspective

历史溯源·津卫之源
History·City

明代 天津三卫疆域及屯堡分布图	清乾隆四年（1739 年） 天津县境图	清光绪二
	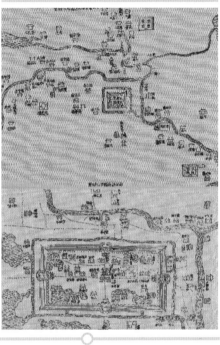	

京杭大运河
（605—610 年）

明朝万历十六年（1588 年），天津兵备道造渡船，在西沽等七地设渡口，这成为天津地区最早的一批官渡。

清朝，西沽渡口已不能满足需求。康熙五十四年（1715 年），官府斥资建造西沽浮桥，是为天津的第一座浮桥。

清朝同治九年（1870 年）十一月，在扩建机器局（东局子）制造军火武器的同时，在西沽择地 350 亩建造火药库，于光绪二年（1876 年）落成，称为"西沽武库"。光绪二十六年（1900 年）六月"庚子之役"中，武库被侵略者炸成废墟。

历史溯源·万商之源
History·Business

盛极津沽

先有估衣街　后有天津卫

津门万商发源地　沽上第一街

金招牌　银招牌　银子窝

有客不打烊

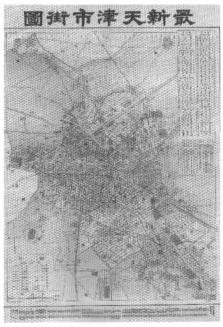

| 0年) | 1946年 天津地图 | 1955年 天津市地图 |

清朝光绪年间，大清河尾闾西退，西沽段河道被泥沙淤涸，浮桥遂废。

清末民初时期，西方文化融入西沽。光绪三十二年（1906年），建立著名西沽教堂。1906—1918年，西沽成了名副其实的基督教公理会活动中心。

21世纪初，此片区开始衰落。

历史溯源·海河之源
History·The source of Haihe River

1918年，三岔河口裁弯取直，西于庄地区成为天津境内海河的源头。

In 1918, the mouth of the Three Forks River was straightened, and the Xiyuzhuang area became the source of the Haihe River.

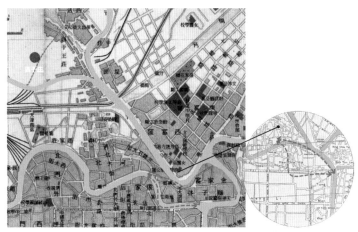

左图 - 清同治九年（1870年）天津地图 | map of Tianjin in 1870
右图 - 1918年天津地图 | map of Tianjin in 1918

限制条件
Restrictions

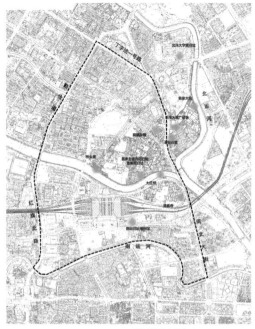

历史建筑分布图 | historical buildings

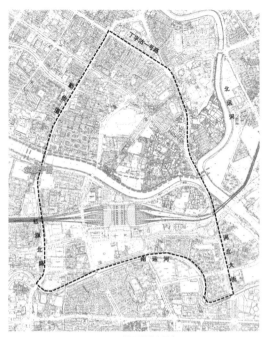

水源干管分布图 | water mains distribution

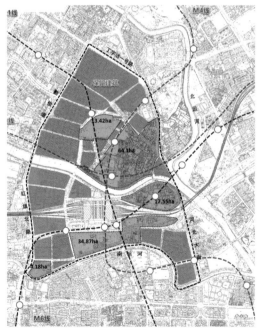

地铁交通网络 | subway transportation network

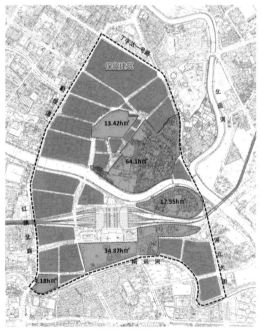

保留地块与新建地块 | reserved plots and new plots

优势条件
Advantage Conditions

区域内高校资源丰富，内有一流学科建设高校，企业能够牵手大学，加速推进科研成果技术转化。

There are many universities, including "project 211" universities and first-class colleges. Cooperating with these universities will accelerate the boost of scientific research.

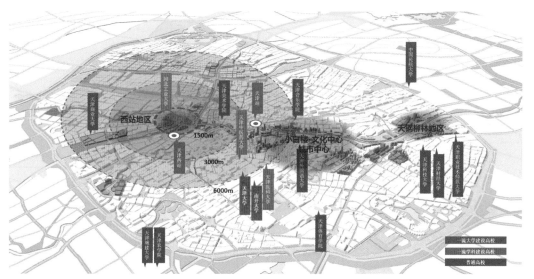

场地条件 | venue conditions

河北工业大学 | Hebei University of Technology

天津美术学院 | Tianjin Academy of Fine Arts

天津商业大学 | Tianjin Business University

天津大学 | Tianjin University

设计原则
Design Principles

总体规划要尊重实际，结合新发展理念，研判西于庄地区的发展逻辑，发现并培育符合区域发展实际的内生动力和外在机遇。中央商务区的概念，源于 1923 年的美国。中央商务区是城市经济、科技、文化、商业高度集聚而形成的城市经济中枢。反观西于庄地区，由于周围路网已经建成，东西干道大都是过境道，南北联系弱，周边区域商业、商务资源稀少，人口老龄化且受教育水平低，距离形成中央商务区的条件有很大差距。规划定位正在由中央商务区向中央活动区转变，实现多活动路径网络的升级转型。

于是围绕该地区建设天津西部 CBD 成为开发目标，最终 2014 年的方案将开发规模放在首位考虑，为政府提供了满足政府、开发企业、居民搬迁各方诉求的折中方案。

The master plan needs to be consider for the reality. We should study the development logic of the area by combining new idea and discover opportunities.The concept of CBD, which originated in the United States in 1923, means the economic, scientific, cultural, and commercial center of the city.The Xiyuzhuang area has a weak connection between the north and the south. Few commercial resources in the surrounding area can be found because the surrounding road network has been built. The aging population is a serious problem, the education level is low. Therefore, there is a big gap for the Xiyuzhuang area to become a CBD area. Recently, The planning and positioning are changed from CBD to CAZ to realize the upgrading and transformation of CAT.

Thus, the construction of the western Tianjin CBD became the development target. The 2014 proposal put the scale of development in the first place, providing the government with a solution to meet the demands of enterprises and residents.

中央商务区
CENTRAL BUSINESS DISTRICT
CBD
商务办公为主
金融贸易 + 信息保险以及相关咨询

- 拥有金融等高盈利行业和机构
- 拥有高级的零售、酒店、文化娱乐等服务
- 建筑容积率、建筑密度、建筑科技水准都是较高的，因此商务空间的集聚度高
- 城市公交和步行网络交通系统发达，交通量大
- 商务区的土地批租价、楼面租金价和楼面出售价高
- 职住不融合，昼夜人口数量差距大

城市跻身为全球供应链辐射腹地发展的门户性锚点 整合区域、城市国际性竞争资源的核心区

"7×24"的活动模式

为了保证街区日夜的热闹氛围，功能必须小规模最大限度地融合基本需求（饮食、教育、医疗等）与其他需求（看展览、看演出、购物等），还必须考虑到季节性问题。

09:00 – 15:00
商务办公活跃时间

15:00 – 19:00
文化休闲活跃时间

19:00 – 23:00
商业购物活跃时间

23:00 – 次日 09:00
酒店居住活跃时间

中央活动区
CENTRAL ACTIVITY ZONE
CAZ

多功能活动的中心
包括商业、居住、企业办公、休闲、餐饮、社交、娱乐、教育、保健、专业服务、小型医疗、政府服务空间以及现代型居住社区

- 高度集聚、链条完整的新兴经济产业
- 面向全球的金融、商务和会展中心
- 多元化高品质的生活配套
- 富有活力的开放空间和生态环境(窄路密网+绿色交通)
- 可经营、可传承的多元文化
- 快速便捷的商旅枢纽

加速培育城市知识创新产业的智慧集聚区 → 全面展现城市美好生活画卷的高活力示范区

多元活动路径网络
CLUSTERED ACTIVITY TRAILS
CAT

 不是仅有孤立的功能单一的中央商务区

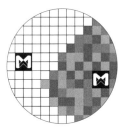

 不是仅聚集于一处的混合功能中心

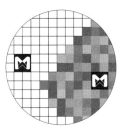

 而是构建一个承载多元功能的网络，串联公共交通节点，依托慢行交通路线，从而均匀散布在城市中，形成一个骨架活力点脉络。

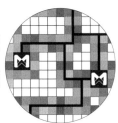

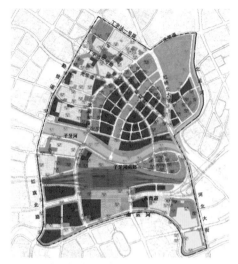

2008 年副中心城市设计 | sub-center urban design in 2008

2012 年副中心城市设计 | sub-center urban design in 2012

2008 年规划：定位为以交通枢纽为发展引擎，凸显水岸特色、天津风情，打造现代、高效、生态型城市副中心。

2012 年规划：定位为以城市综合交通枢纽为依托的区域型生产性服务业中心，是面向天津西北的商贸中心。

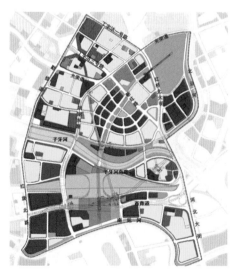

2014 年副中心城市设计 | sub-center urban design in 2014

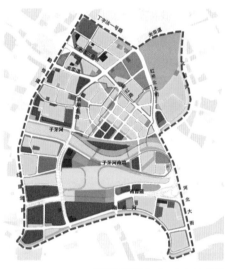

2018 城市设计 | urban design in 2018

2014 年规划：定位为综合性城市副中心核心组成部分，赋予其城市中心职能，使之成为新的生产服务、生态宜居中心。

2018 年规划：定位调整为以智慧创新为核心，以本土文化为依托，以生态保育为本底，打造智慧、人文、生态相得益彰的新经济活力区。

西于庄近期规划总平面图 | recent general plan

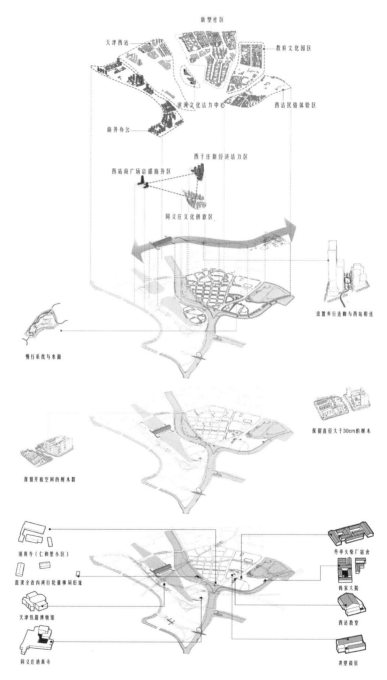

爆炸分析图 | exploded views

设计策略
Design Strategies

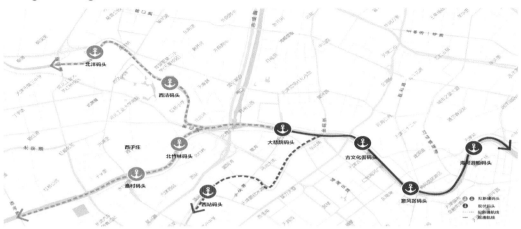

运河复航规划图 | canal resumption plan

延续历史文脉，重塑探索之城。

运河复航：将北运河、子牙河恢复通航（公交化），与海河游船航线相衔接，增强游客对天津河运文化魅力的体验。

"最天津"文旅体验区：以水、陆双重游览路线串联沿河各要素，强化游客对天津本土文化魅力的体验。

Protect the historical lineage and build the city of exploration.

Resume the canal: Reopen the North Canal and Ziya River public transportation. Connect them with the Haihe River to optimize the experience of the trip to Tianjin for the visitors.

"The Tianjin Style" cultural tourism area: Two tour routes, one by water and one by land, build the attractions along the river to optimize visitors' experience of Tianjin.

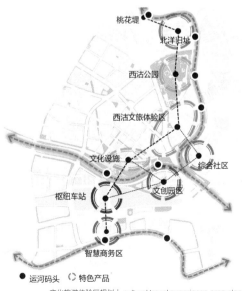

文化旅游体验区规划 | cultural travel experience area planning

现存文化旅游建筑 | existing cultural brigade buildings

依托周边城市主干道路网，在提升周边现状居住环境的前提下，联通外部交通，优化内部路网结构，增加两处跨子牙河通道形成了四横四纵、开放网格式的路网格局。

According to the road network, we connect the external road system to the internal road network structure in the purpose of improving the living environment. We have added 2 bridges across the Ziya River, forming a open road network pattern of four horizontal avenues and four lengthwise avenues.

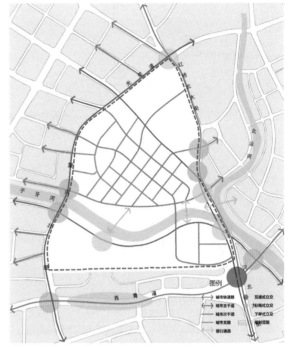

现存文化旅游建筑 | existing cultural buildings

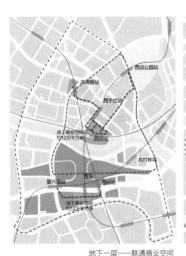

地下一层——联通商业空间
the 1st basement -connects the commercial space

地下二层——停车空间
the 2nd basement— parking space

地下三层和四层——换乘空间
the 3rd and 4th basement — transfer space

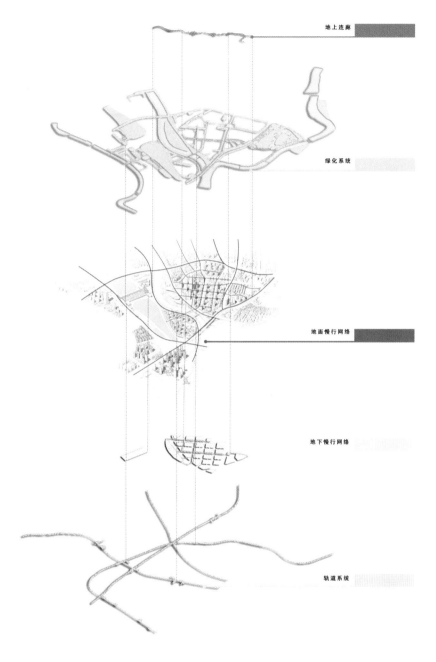

西于庄 TOD 爆炸分析图 | analysis of TOD

慢行系统
Slow Traffic System

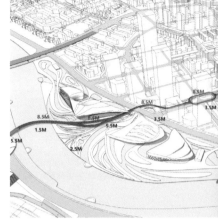

新规划提出，商住比由8：2修正为6：4，减少商业商务办公空间，增加居住和混合业态空间，保证区域内24小时的活力繁荣。居住区增加商业配套设施比例，总体控制在10%左右，保证5分钟、10分钟、15分钟生活圈的便利与活力。通过联通西站后广场、建设步行桥使西于庄可以直接承接西站的客流，同时设置绿轴直达西沽公园，使城市慢行道联网成片。

The new plan proposes that the ratio of commercial to residential is amended from 2:8 to 4:6, reducing business offices and increasing residential and mixed businesses to ensure 24-hour space vitality. The overall proportion is about 10% to ensure the convenience and vitality of the 5-10-15-minutes living circle. Cooperating with the development of the Tianjin West Railway Station square, we build a pedestrian bridge so that Xiyuzhuang can provide a place for the passengers from the Tianjin West Railway Station. At the same time, they can go to the Xigu Park through the green space. In which way, the urban slow walking system is interconnected into a whole network.

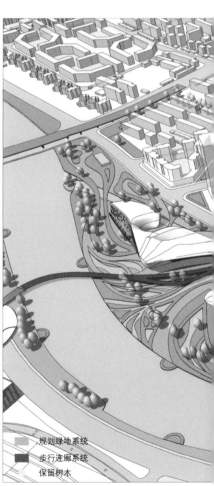

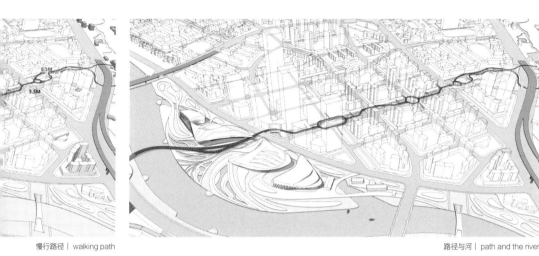

慢行路径 | walking path　　　　　　　　　路径与河 | path and the river

慢行系统整体规划 | slow traffic plan

西于庄形成、发展的百年当中，仍葆有自己的荣耀、自己的乡愁、自己的民俗民风。

In the hundreds years of development, Xiyuzhuang has reserved its own glory, feature and culture.

鸟瞰图 | aerial view

西开教堂位于天津市和平区滨江道独山路,原来墙子河外老西开一带,即现在营口道与西宁道交口旁,坐西南朝东北。1916 年由法国传教士杜保禄修建,地处法租界,又名法国教堂。

教堂采用法国罗曼式建筑造型,高 45 m,建筑面积 1 585 m²,平面呈拉丁十字式,正面入口两座塔楼,中部一座主塔呈"品"字形,墙身由红、黄钢砖砌筑,屋顶木结构表面覆盖的铜皮在漫长的岁月中变成绿色。内部采用通廊式布局,左右各 7 根立柱,墙面彩绘壁画华丽大气,这是当时华北地区规模最大的一座教堂。教堂附近陆续开办过西开小学、若瑟小学、圣功小学、若瑟会修女院法汉学院(今天津市第二十一中学)、天主教医院(天津市中心妇产医院的前身),形成一大片教会建筑群。2008 年西开教堂被天津市政府列为

西开教堂 Xikai Church

用地面积：2.9 hm²
建筑规模：24 000 m²

城市设计团队：
天津市城市规划设计研究总院

修建性详细规划团队：
天津市城市规划设计研究总院

获奖情况：
2015 年度天津市优秀城乡规划设计奖二等奖

鸟瞰图 | aerial view

区位图 | location

重点保护等级历史风貌建筑。

正如威廉·莫里斯所言，"它（建筑）从任何意义上都不是我们任意处置的对象，从某种程度上来说我们只不过是代管人而已。"但我们作为代管人，却没有管理好前人留下的"瑰宝"，西开教堂地区用地由于历史原因被不断压缩，周边环境恶劣，其应有氛围丧失，已陷入拥挤混乱之中，对城市而言将一个文物放在一个杂乱无章的区域，使城市文化沦落为街道边的西洋景是何等无奈。因此，重新梳理教堂周边区域的空间环境，修缮西开教堂及附属建筑，提升其空间品质，发挥其文化价值显得日益急迫。

The Xikai Church is in Dushan Road, Binjiang Street, Heping District, Tianjin. The original wall outside the old Xikai area is where near the current Yingkou Road and Xining Road, next to the intersection of southwest facing northeast. The French missionary Paul-Marie Dumond presided over the construction in the French Concession. The Xikai Church is also known as the French Church.

The church adopts French Romanesque style, 45 meters high, with a building area of 1585 square meters, a Latin cross-shaped plan, two towers at the front entrance, a main tower in the middle. Walls made of red and yellow steel bricks. Roof wooden structure covered with copper turns green with ages. The interior structure is connected by the hall and the corridor, with 7 columns on the left and right. The walls inside are decorated with gorgeous painting. It was one of the largest churches in Northern China. In 2008, it was listed by Tianjin Government as one of the historical buildings for protection.

Xikai church is the treasure of Tianjin. As William Morris said, "It is never a thing that we can take charge of, to some point, we are just a keeper." However, as a keeper, we did a bad job. The area of Xikai Church has been compressed due to historical reasons. The environment around has deteriorated and lost its feature and became a crowded and chaotic place, which is a pity for the city to put a cultural relic in a cluttered area and leave it alone on the street. Therefore, it is urgent to reorganize the spatial environment of the area around the church, repair the Xikai Church and buildings around, improve its quality, and bring its cultural value into play.

西开教堂周边建筑的遮挡 | sight shield around Xikai Chuch

教堂被现代建筑重重包围 | surrounded by modern architectures

历史
History

近代天津曾经修建过百余座教堂,这些遍布津城的教堂,见证了天津教案、义和团运动、老西开事件等中国近代史上的著名事件。

天津现存的诸多教堂中有的难以形成保护片区文化带,有的已被损毁破坏无法使用,只有西开教堂是目前国内现存规模最大的教堂,迄今仍在使用,并且具有极其重要的宗教、历史和旅游价值。

西开教堂自1916年建成以来,历经百年沧桑几度严重受损,后经修复完善,成为天津乃至华北地区最大的天主教堂,是天津天主教会的中心。

Hundreds of churches were built in Tianjin in modern times.These churches, which are located all over the city, have been through famous events in China's modern history, such as the Tianjin Church Case, the Boxer Rebellion Movement, and the Old Xikai Incident.

Some of the existing churches in Tianjin have been destroyed and are no longer in use, but the Xikai Church is the only largest existing church in China, which is still in use and has extremely important religious, commercial, and tourist values.

Since it was built in 1916, the Xikai Church has been severely damaged several times in last century. Now, it has been restored and improved to become the largest Catholic Church in Tianjin and even in North China.

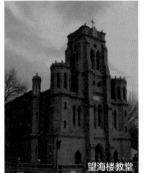
望海楼教堂

紫竹林教堂

安里甘教堂

富辛庄教堂

犹太教堂

仓门口教堂

冈纬路教堂

1916年建成 | built in 1916

20世纪20年代 | 1920s

20世纪30年代 | 1930s

1939年水灾 | 1939 floods

20世纪50年代 | 1950s

20世纪60年代 | 1960s

20世纪70年代 | 1970s

20世纪80年代 | 1980s

20世纪90年代 | 1990s

现今 | now

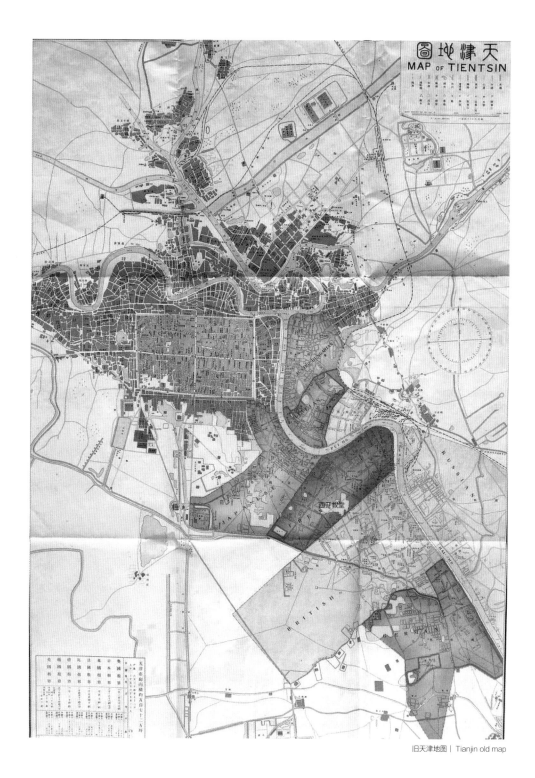

旧天津地图 | Tianjin old map

它一百多年来安静低调地矗立在那里，外墙在岁月的侵蚀下显出斑驳与沧桑，室内高耸的两排柱子将空间分为三个区域，优雅的色调及装饰在巨大的空间中显得温暖、庄重，从靠近、进入到离开让人心生平实的敬意。

当你离开它时，一个不经意的视角吸引了大家。从正门出来向北沿滨江道能够看到天津东站的钟楼！这难道是一条百年前按古典主义手法留下的轴线？经过查阅天津近代城市历史，翻阅九国租界规划，梁思成、张锐的天津愿景，日本人的"大天津规划"等档案资料，我们发现这不是当初的构想，虽然在梁张规划概念中留有一笔，但并未实现。那么，应该是后来天津站建设选址时，设计师有意而为，因为西开教堂是不会动了，只有新建的后来者来适应前者，在空间视线上实现百年间的守望，联想至此不禁让人对天津站的设计者肃然起敬。

It has existed here for a hundred-year, and its delicate facade has become mottled over the years. Inside, two rows of columns divide the space into three areas. Elegant tones and decorations reveal a warm, refined atmosphere in the vast space. From approaching, entering to leaving people will feel peace and be infected by the divine power.

When leaving the church, we found a special view. Along the northeast along Binjiang Road, we were able to see the clock tower of Tianjin Railway Station. Could this be a design left by the classical approach a century ago? After going through the modern city history of Tianjin, the planning of the Nine Kingdoms Concession and the Japanese planning of Greater Tianjin, we found that this was not mentioned. Although it was mentioned in Liang Zhang's planning concept, it was not realized. Then, it should be the idea of the designer when the Tianjin Railway Station was built. Xikai will not change, but the new construction of new buildings can adapt. Because of the spatial sightline, we feel great admiration for the designers of the Tianjin Railway Station.

区位图 | location

不同历史建筑的独特性在于它们在社会、历史、艺术审美、自然、人的精神等深层次存在的价值,也在于它们与物质、视觉、精神的以及其他文化层面的环境背景产生的联系。此外,周边环境也是重要资源。因此从环境入手,对历史建筑开启保护工作至关重要。

The uniqueness of different ancient buildings lies in their value with social, spiritual, historical, artistic, natural, and other aspects. Besides, its connection with the physical, visual, spiritual, and other cultural aspects also change its feature. The surrounding environment are also essential resources. Starting from the environment, we started the conservation work.

教堂内部 | inside the church

教堂内部 | inside the church

三版规划
Three Edition Plan

十几年来，西开教堂地区的规划一共做过三版，分别是不同历史时期背景下的构想。第一版为2003—2005年的规划；第二版为2006—2008年的规划；第三版为2009—2013年的规划。

For more than ten years, three versions of the Xikai church area plan were made, each with a different background. The first edition is from 2003-2005; the second edition is from 2006-2008; the third edition is from 2009-2013.

2003—2005年规划方案 |
planning scheme for 2003-2005

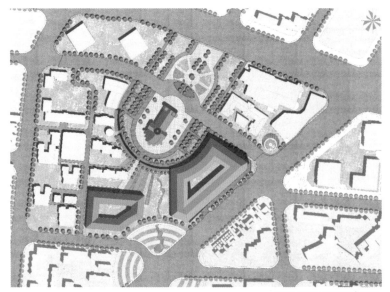

2003—2005年规划方案总平面 |
general plan for 2003-2005

2006—2008年规划方案｜
planning scheme for 2006-2008

2009—2013年规划方案｜
planning scheme for 2009-2013

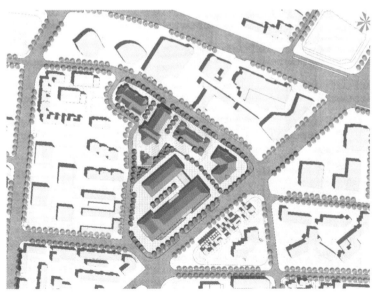

2006—2008年规划方案总平面｜
general plan for 2006-2008

2009—2013年规划方案总平面｜
general plan for 2009-2013

三版规划：第一版
Three Edition Plan: The First Edition

2003—2005年的规划，拆除国际商场等周边建筑，以教堂为核心，用它来组织空间充分体现其价值，这是一个比较理想的规划。

2003-2005 plan proposed to demolish the international shopping mall and use the church to dominate the space, which is an ideal way of planning.

2003—2005 年规划总图 | general plan for 2003-2005

2003—2005 年方案鸟瞰图 | aerial view of 2003-2005 planning scheme

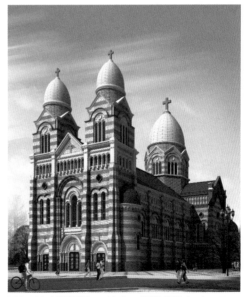

2003—2005 年规划方案效果图 | effect picture of 2003-2005 planning scheme

2003—2005 年规划局部 | partial effect of 2003-2005 planning scheme

三版规划：第二版
Three Edition Plan: The Second Edition

2006—2008 年的规划，拆除东、南、西三侧区域的建筑，保留北侧国际商场，这是一个向现实妥协的规划。

2006-2008 plan proposed to demolish the buildings on the east, south, and west sides and keep the international shopping mall on the north side, which is a realistic plan.

2006—2008 年规划方案效果图｜effect picture of 2006-2008 planning scheme

2006—2008 年规划方案总平面｜general plan for 2006-2008

2006—2008 年方案鸟瞰图｜aerial view of 2006-2008 planning scheme

2006—2008 年规划方案局部效果｜
partical effect of 2006-2008 planning scheme

三版规划：第三版
Three Edition Plan: The Third Edition

2009—2013年规划，拆除北侧国际商场配楼，打开视线通廊，东、南、西三面拆除局部治理的方案，是已经均衡协调各方利益保护规划。现规划在西宁道南侧已经完成，但北侧还未落实。

保护传统建筑是一项复杂的工作，只有让它们留下来并予以保护，才能真正提升城市的文化价值。

历史和文化建筑在现实环境中历经沧桑，可能人们想象中的那个世界，繁华褪去已经显现颓势。而西开教堂就像蒙尘的宝石，等待灰尘被除去，重现光芒照亮周边，将天津的精彩故事讲述下去。天津是一座有故事的城市，传承天津的历史文化是我们每个市民的责任和使命。

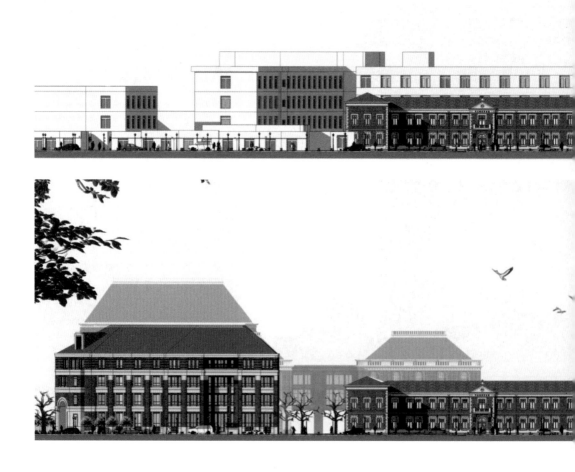

2009-2013 plan proposed to demolish the podium of the international shopping mall on the north side to open up the sightline. Now the plan has been completed on the south side of Xining Road, but the north side is not yet under construction.

Protect traditional buildings and activate their value. In that way we can enhance the cultural value of the city.

People are so obsessed with rapid development that history and culture cannot be passed on. Such prosperity can only be maintained for a short time. The potential of Xikai Church is waiting to be developed so that the beauty of Tianjin will be remembered. The Xikai Church project may be an opportunity to help people to look back and carry out the plan. Tianjin is a city with abundant stories, and each of citizen has the responsibility to pass it on.

第三版方案西开教堂改造前 | the third edition plan III –before the transformation

第三版方案西开教堂改造后 | edition plan III –after the transformation

在天津城市发展的历史中,"津味"文化一直是这座城市的灵魂,它是引领城市永续发展的内驱力。沿海河分布的原九国租界对天津城市结构影响深远,至今我们仍旧能强烈地感受到这里华洋杂糅的气息。其中,原英租界即现在的西康路至海河边的五大道和小白楼区域,是保存最完好、特色最鲜明的英式街区。由于南京路的建设,原英租界被分为两个部分,2009 年市政府为恢复历史街区,保护文化遗产,决定启动搬迁工作,同时进行以泰安道解放北园(原维多利亚公园)为核心的城市更新建设。

泰安道五大院区域沿海河向北,经解放北路,可与津湾广场相连,向西南可与小白楼商圈相接,直至五大道。新五大院的建设开发可以将有"北方华尔街"之称的解放北路与小白楼商圈及五大道街区建立直接联系,形成沿海河的历史片区,加上天津津湾广场的建成为区域经济发展、历史街区复兴带来巨大机会。

泰安道五大院
Five Courtyards in Tai'an Road

用地面积：49.5 hm²
建筑规模：404 000 m²

城市设计团队：
天津市城市规划设计研究总院

修建性详细规划团队：
天津市城市规划设计研究总院

合作团队：
天津市城市规划设计研究总院
天津市博风建筑工程设计有限公司
天津大地天方建筑设计有限公司
天津华汇工程建筑设计有限公司

获奖情况：
2019年度全国优秀勘察设计项目

鸟瞰图 | aerial view

区位图 | location

五大院地区有16处历史建筑遗迹，像宝石一样散落在地块之中。如何保护和利用这些分散的资源，规划师提出重塑街区、连接空间的方案，将既有建筑重新组织，在以其存在为基础构架的街区空间中，通过强调贴线率、形塑完整街道，并通过开放院落空间之间的连接，形成院落群之间有机的步行网络系统，从而从人的角度出发恢复街区活力。

泰安道五大院每个院落所处的环境各不相同，在这样严苛的外部条件下，设计既要有统一的原则来指导具体的设计工作，又要给各个院落的"主创设计师"足够的空间，发挥积极性、创造性，解决每个院落的特殊问题。

During Tianjin's urban development, the culture of Tianjin has always been the soul of the city, and it is the driving force that leads Tianjin to sustainable development. The concession distributed along the river has had a profound impact on the urban structure of Tianjin, and the mix of Chinese and foreign culture still affects today's lifestyle. The British Concession, which is now from Xikang Road to the Five Avenues and Xiaobailou area along the Haihe River, is the most well-preserved neighborhood with distinctive British features. Due to the construction of Nanjing Road, it is divided into two parts. In 2009, the government launched to relocate the area and restore the historic district to protect the cultural heritage. Besides, the urban renewal construction was carried out with Victoria Garden on Taian Road as the core.

The area can be connected to Jinwan Square along the river to the north, via Jiefang North Road, and to the southwest to the Xiaobailou shopping district up to the Five Great Avenues. The construction and development of the new Five Great Courtyards can establish a direct connection between North Jiefang Road, known as the "Wall Street of the Northern China", and the Xiaobailou shopping district and the Wudadao neighborhood, forming a historical district

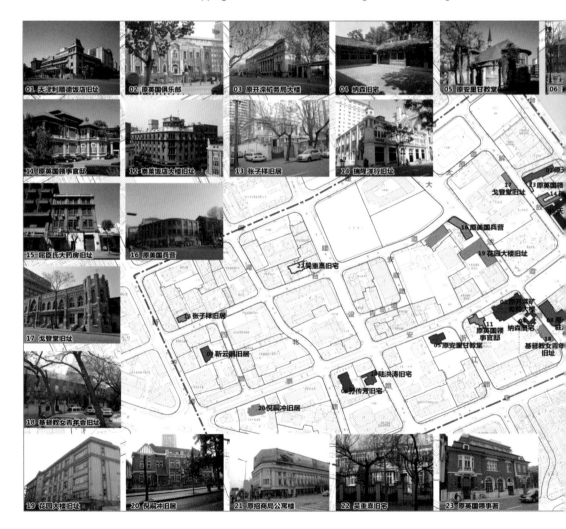

along the river. Together with the completion of Tianjin Jinwan Square, there is a huge opportunity and potential for regional economic development and revival of the historic district. There are 16 historical relics in the Five Courtyards area, dropped on the site. To use and preserve these resources, the planners proposed an approach to reshape the neighborhood and link the spaces. The existing buildings are reorganized in neighborhood planning by emphasizing the rate of roads' lines, shaping streets, and opening up the connections between courtyards. The neighborhoods are revitalized by building up the pedestrian system between different courtyards.

Each of the five courtyards is in different conditions. Under such conditions, we must have common principles to guide the specific design work, also give each courtyard's designer-in-chief enough freedom to create and solve their unique problems.

《道历史文化街区保护控制性详细规划》划定的保护建筑 | protected buildings

泰安道历史街区中的利顺德大饭店 | The Astor Hotel

1860 年英租界的原定界线 | British Concession in 1860

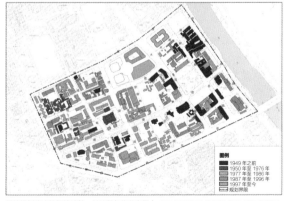

保护规划确定的保护建筑，其建筑年代分布图 | chronological distribution of buildings

189

设计导则分为两个部分，一是关于空间、街道的原则，提出：重塑街区、连接系统、保护情境，坚守匠心；二是平衡开发量的原则，提出：以解放北园为中心（经研究历史资料信息及进行日照分析后决定）将东、南、西三侧的建筑高度控制在 24 m 以下，四号院高度控制在 40 m 以下，其余建筑的开发量用五号院高层建筑来平衡。

The urban design principles include two parts. The first is the principle of space and streets. It is about reshaping the neighborhood relation, linking the systems, and protecting the context. The second is the principle of balancing the development volume. By centering on Victoria Garden, we study the information of historical materials and conduct sunlight analysis. The heights of the building in the east, south, and west should be under 24 meters. The elevation of the NO.4 courtyard should be under 40 meters. Besides, the rest should compare to the highest building in the No.5 oourtyard.

鸟瞰图 | aerial view

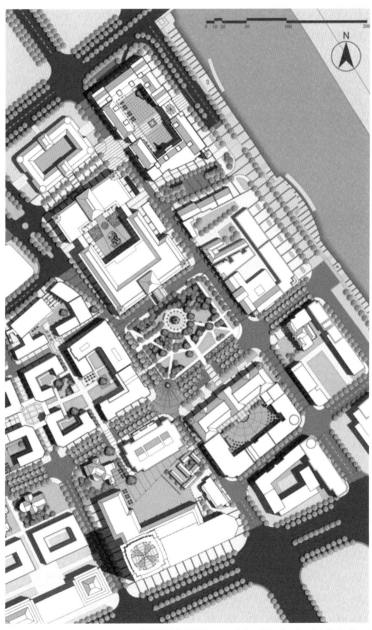

泰安道五大院总平面 | general plan of Five Courtyards in Tai'an Road

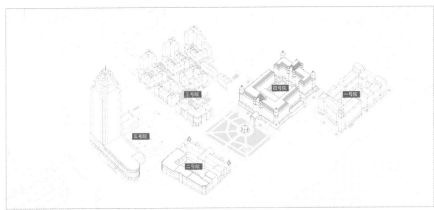

院落空间分析 | courtyard space analysis

鸟瞰图 | aerial view

泰安道实景图 | scene of Tai'an Road

193

一号院在各个院落中所受限制最少，但其紧临海河南侧，区域内有利顺德大饭店，从海河沿线来看，该院落对区域风貌的影响最大。设计师在南部通过设置钟塔，给五大院地区在海河上建立一个标志，同时在东西两侧设计有宽大的门洞，沿海河的景观可透过内院从室内看到，同时建筑内部也可通过门洞融入城市街景中。

一号院实景图 | scene of No.1 courtyard

The No. 1 courtyard, which is next to the southern side of the Haihe river has the least restricts but has the greatest impact on the regional landscape from the view along the Haihe River. The designers created a symbol for the Five Courtyard area by setting up a clock tower in the south in the courtyard. Besides, two large-scale arch gates are designed on the east and west sides. In this way, people can see the landscape along Haihe River through gate and within the inner space. The inner space in the buildings is also involved in the city landscape through the doorways.

细部图 | detail

一号院实景图 | scene of No.1 courtyard

一号院入口 | entrance of No.1 courtyard

二号院内有天津第一饭店、原十八集团军驻津办事处及屈臣氏大药房旧址三处历史建筑，加上多年来建筑内部改扩建多处，状况复杂，设计师不仅需在空间形象上下功夫，同时需要在满足功能的条件下对设备及地下空间进行重新梳理，实现"内外兼修"，使得二号院外部造型严整、新旧部分协调，内部功能完备，各类设施完善。

The No.2 courtyard includes three cultural relics: the First Hotel, the Eighteenth Route Army building, and the Youth Federation square. There are changes and expansions inside the building over the years. The current conditions are complex, so the designers need to work not only on the outside space but also to reorganize the equipment and underground space to meet the function. The aim is to construct a modern external shape in harmony with the historic buildings and to integrate the internal facilities.

二号院实景图 | scene of No.2 courtyard

立面细部 | detail of facade

屋顶细部 | detail of roof

立面细部 | detail of facade

三号院在解放北园西侧，与利顺德饭大店相对，它的居住区的定位及原美国兵营的存在，使设计师决定采用"一大带四小"的院落布局，内部道路通而不透，院落间利用高差形成从城市公共区域到半私密区域、私密区域的秩序，在地块的四个方向上均留有通道，保证步行的连续顺畅。

The No. 3 courtyard is opposite the Astor Hotel on the west side of the Victoria Garden. It is positioned as a residential area. Because of the residential area and the American Barracks, the courtyard consists of one large plot with four small plots. The internal roads that only let people get through. Different heights between the courtyards form the spaces from the urban public area to the semi-private area and private areas. The passages in all four directions of the plot ensure a continuous and smooth walk.

三号院咖啡厅 | cafe in No.3 courtyard

三号院实景图 | scene of No.3 courtyard

三号院街角 | street corner of No.3 courtyard

鸟瞰图 | aerial view

四号院实景图 | scene of No.4 courtyard

四号院在解放北园南侧，原址戈登堂的大部分已毁于地震，仅在区域东侧遗存一个小建筑，保留有戈登堂的旧迹。由于同二号院相对，内院采用对称方式进行空间呼应，同时通过设置出入口，强化轴线的存在。四号院内院东西两侧各留有门洞，与一、二号院相连，四号院内院门洞同一号院钟楼形成对景。

The No.4 courtyard is on the south side of Victoria Garden. Most parts of the original Gordon Hall in the yard have been destroyed by the earthquake, leaving only a few on the east side of the base. As it is opposite to the No. 2 courtyard, the inner courtyard adopts a symmetrical layout to match with it. At the same time, we reinforce the existence of the axis by setting up entrances and exits to the plot. There are doorways on the east and west sides of the inner yard of No. 4 courtyard, which are connected with No. 1 and No. 2 courtyards. Besides, the gate of the inner yard of No. 4 courtyard forms a counterpoint to the clock tower of No. 1 courtyard.

四号院入口 | entrance of No.4 courtyard

四号院鸟瞰图 | aerial view of No.4 courtyard

解放北园是泰安道地区历史最悠久的存在，这也许应该归功于134年时光的积淀。解放北园始建于1860年，作为天津首座租界公园，在规划、建筑、植物造型等方面都具有折中主义风格。我们打破"具体化"的想法，在园中做了一些尝试，保护了场地内的所有植物，坚持传统布局，以四条辐射状道路连接四个角门，保持随意中蕴含规律的气质，用最简单的方式为人们提供零距离接触历史文化的机会。

五号院在三号院南侧，解放北园西南角是新旧界面过渡的位置。场地内有原开滦矿务局大楼、纳森旧宅、基督教女青年会旧址，设计沿用与其他院落相同的处理方式，新旧建筑采用同一种更平和松散的风格，由于既有建筑差异巨大，采用"对话方式"组建成一个围合的院落，使老建筑的环境氛围得以保留。内向复古、外向时尚的风格以及高层的设置，将时空的距离拉长。在文化方面的包容性，通过新老建筑的对话充分体现。

以现代技术改良砖砌工艺，泰安道五大院的英式砖楼是天津的地标性建筑。天津人无论走到哪里，记忆中的旧时光总会萦绕在脑海，海河风貌与河岸的庭院将永远印刻于心。时光荏苒，岁月变迁，天津一直在那里，守候着游子的归来。

Victoria Garden is the most attractive place in the Tai'an Road area, which could probably be attributed to its 126 years history. Victoria Garden was built in 1860 in the concession in Tianjin, with an eclectic style of planning, architecture, and planting. We did not draw a figure but made some experiments in the garden, protecting all the plants on the site.Sticking to the traditional layout, we made four radial roads connect to the four corner gates.It is the simplest way to provide people with the opportunity to be close to its history and culture.

The No.5 courtyard is in the south of the No.3 courtyard, and the southwest corner of the Victoria Garden locates between the old and new interfaces. The Kailuan Mining Bureau, the former residence of Nathan, and the Women's Union are all on the site. The design follows the style of the other courtyards, being more peaceful and loose for the old and new buildings. Due to the differences between the existing buildings and the new ones, we combined different elements of them and built a new enclosed courtyard. In which way the ambiance of the old buildings is preserved. The inside vintage-looking and outside modern-looking with high buildings provide a different space for people to experience various times and space. The tolerance of the cultural dimension is reflected through the dialogue between the old and the new.

The English-style buildings made of brick are improved with modern techniques. Using these bricks, the Five courtyards will become the new city's iconic places. Wherever people go, these memories of the old days will stay with them. The culture of the Haihe River and the courtyards along the riverbank will always be in their memories. Tianjin will always be there for people to come back.

解放北园 | Victoria Garden

公园景观 | park landscape

公园小径 | path in the park

公园沿街一侧 | park view along the road

鸟瞰图 | aerial view

设计团队名单 | Team Members

天津市城市规划设计研究总院建筑设计一院

建筑设计一院现有员工100余人，其中国家一级注册建筑师、注册工程师30余人。近年来，建筑设计一院依托总院优势，实现了"全过程设计""全流程咨询""全方位把关"的全产业链模式，推进重大公共项目落地。提供从城市设计、前期定位、开发策划、规划、建筑、环境景观等全方位、全流程的工程解决方案。同时，推行"总师责任制"，将先进的设计理念及管理模式贯穿始终，保证设计思想的高质量落地。

建筑一院团队成员（按姓氏拼音排序）

安志红、陈伟杰、陈旭、崔栋、崔磊、邓超、狄阙、董天杰、窦金楠、高艳、宫田芮、郭金未、郭鹏、郭宇、韩海雷、韩佳昕、韩立强、韩宁、侯勇军、胡婷婷、黄波、黄轩、姜雪、解广锋、金彪、康程杰、类成诚、李超、李传刚、李津澜、李钧、李丽媛、李显、李欣艳、李兴宇、李绪良、李雪256、李雅洁、刘畅、刘德忠、刘建红、刘磊、刘亮、刘瑞平、刘月、卢洪权、毛晓亮、孟沫、莫旭昭、牛筱甜、潘宏达、潘林、齐思彤、邱雨斯、曲小美、饶晖、任艳琴、佘江宁、沈伦、石光、苏渊、孙科章、田明、田轶凡、万雪梅、王斌、王立剑、王连顺、王文昌、王洋、王智聪、吴宏婷、吴书驰、吴振兴、邢哲、徐礴骁、闫菁清、闫艺、殷利、袁征、张江铭、张宁、张润兴、张愈芳、赵彬、赵春水、赵堃、郑秀全、周娜

天津文化中心
师武军、赵春水、侯勇军、崔磊、陈旭、田明、潘福超、张小平、张潇、魏星、白惠艳、李艳蓬

天津滨海文化中心
赵春水、董天杰、田垠、侯勇军、陈旭、陆伟伟、韩宁、安志红、吴书驰、田轶凡、张萌

济宁文化中心
赵春水、侯勇军、韩宁、安志红、陈旭、田轶凡、刘畅、田园、崔磊、阳建华、饶辉、刘津雅、类成诚、李绪良、高艳、刘瑞平、史永奇、莫旭昭、庄子玉、李娜、费边·韦瑟（Fabian Wieser）、朱胜利、高学浩、张峻

双青新家园
赵春水、董天杰、陈旭、邱雨斯、田园、宫媛、赵光、郑兆唯

义乌异国风情街
赵春水、田垠、吴书驰、赵彬、李超、石光、齐思彤、陈旭、李传刚

河西八大里
赵春水、董天杰、马松、陈旭、郭宇、刘瑞平、石光、高媛、魏星、张潇

天津市第一热电厂
赵春水、董天杰、李津澜、邱雨斯、刘瑞平、田轶凡、高瑾、李薇

西站西于庄
赵春水、吴书驰、陆伟伟、李超、李传刚、齐思彤、佘江宁、陈伟杰、张泽鑫

西开教堂
赵春水、张润兴、韩海雷、刘畅、黄轩、张愈芳、赵彬、崔磊、谢瑞建、顾菲菲、袁悦

泰安道五大院
黄晶涛、赵春水、马健、王超、赵庆东、张润兴、马凯、张楠楠、田园、田垠

照片拍摄者（按姓氏拼音排序）
关永辉、郭鹏、郭晓音、李勇、田园、魏刚、邢哲、战长恒、张明贺、张胜强、甄琦

照片提供单位
天津市规划和自然资源局、天津市滨海新区文化中心投资管理有限公司、天津市建筑设计研究院有限公司、上海兰斯凯普城市景观设计有限公司、济宁城投控股集团有限公司、天津天房酒店管理有限公司丽思卡尔顿分公司

图纸绘制者（按姓氏拼音排序）
陈伟杰、陈旭、崔磊、狄阙、宫田芮、郭宇、黄轩、康程杰、李传刚、李雅洁、刘畅、刘瑞平、石光、王智聪、吴书驰、薛腾、闫菁清、闫艺、张萌、赵彬、周娜

项目资料提供者（按姓氏拼音排序）
陈伟杰、陈旭、崔磊、郭宇、韩海雷、黄轩、李超、李传刚、李津澜、刘畅、刘瑞平、邱雨斯、石光、田轶凡、田园、吴书驰、张萌、赵彬、周娜